Adsorption on and Surface Chemistry of Hydroxyapatite

Adsorption on and Surface Chemistry of Hydroxyapatite

Edited by
Dwarika N. Misra

America Dental Association Health Foundation
Research Unit, National Bureau of Standards
Washington, D.C.

Plenum Press • New York and London

Library of Congress Cataloging in Publication Data

Symposium on Adsorption on and Surface Chemistry of Hydroxyapatite (1982: Kansas
City, Mo.)
Adsorption on and surface chemistry of hydroxyapatite.

"Based on proceedings of the Symposium on Adsorption on and Surface Chemistry of
Hydroxyapatite, held September 12–17, 1982, at the ACS meeting in Kansas City,
Missouri"—T.p. verso.
Includes bibliographical references and index.
1. Hydroxyapatite—Congresses. 2. Adsorption—Congresses. 3. Surface chemistry—
Congresses. I. Misra, Dwarika N. II. American Chemical Society. III. Title.
QD181.P1S96 1982   546′.71224   83-24738
ISBN 0-306-41556-X

Based on Proceedings of the Symposium on Adsorption on and Surface
Chemistry of Hydroxyapatite, held September 12-17, 1982, at the
ACS meeting in Kansas City, Missouri

©1984 Plenum Press, New York
A Division of Plenum Publishing Corporation
233 Spring Street, New York, N.Y. 10013

PREFACE

Hydroxyapatite is the structural prototype of the main inorganic constituent of bone and teeth and, together with fluorapatite, is also one of the principal minerals in commercial phosphate ores. The adsorption characteristics and surface chemistry of hydroxyapatite are important in understanding the growth, dissolution and adhesion mechanisms of bone and tooth tissues and in elucidating the factors in mineral beneficiation such as floation and flocculation.

This volume essentially documents the proceedings of the symposium on the same topic held at the American Chemical Society Meeting in Kansas City, MO, September 12-17, 1982. It includes a few papers which were not presented at the symposium but does not comprise the entire program.

This volume provides, on a limited scale, a multidisciplinary overview of current work in the field of adsorptive behavior and surface chemistry of hydroxyapatite and includes certain review articles. There are two papers each on adsorption, adsorption and its effects on crystal growth or dissolution kinetics, effects of electrochemical parameters on solubility and adsorption, and newer physical methods (exoemission and high-resolution NMR) of examining hydroxyapatite surface. There is one paper each on structure modelling of apatite surface based on octacalcium phosphate interface and on biodegradation of sintered hydroxyapatite.

I wish to express my appreciation to the management of the American Dental Association Health Foundation for permitting me to organize the symposium and to edit this volume, acknowledging primarily the understanding and patience of Drs. R. L. Bowen and W. E. Brown. I thank the reviewers for their time and valuable comments and the authors for their enthusiastic cooperation. I am particularly grateful to Mrs. L. E. Setz for her prompt and competent secretarial services throughout the project period, and I especially thank my wife, Chandra, for her cooperation and forbearance during the same period.

August, 1983   Dwarika N. Misra
Gaithersburg, MD

CONTENTS

THE INFLUENCE OF MELLITIC ACID ON

THE CRYSTAL GROWTH OF HYDROXYAPATITE

Zahid Amjad

BFGoodrich Chemical Group
Avon Lake Technical Center
P. O. Box 122
Avon Lake, OH   44012

ABSTRACT

The constant composition method has been used to study the influence of mellitic acid (benzene hexacarboxylic acid) on the kinetics of crystal growth of hydroxyapatite (HAP) at low constant supersaturation. Addition of mellitic acid to the calcium phosphate supersaturated solutions has a striking inhibitory influence upon the rate of crystal growth of HAP. The effect is interpreted in terms of adsorption, following the Langmuir isotherm, of mellitate ions at the active crystal growth sites.

INTRODUCTION

The mineralization of calcium phosphates, which is present in biological hard tissues, is complicated by the existence of four phases namely: dicalcium phosphate dihydrate ($CaHPO_4 \cdot 2H_2O$); octacalcium phosphate ($Ca_8H_2(PO_4)_6 \cdot 5H_2O$, OCP); tricalcium phosphate ($Ca_3(PO_4)_2$, TCP); and hydroxyapatite ($Ca_5(PO_4)_3OH$, HAP). Results of previous studies[1-4] on the spontaneous precipitation of calcium phosphates at physiological conditions indicate that the kinetically favored precursor phases are formed prior to the formation of the thermodynamically stable HAP. Moreover, it has also been suggested that such precursor phases can be stabilized by the presence of certain additives.[5,6]

The influence of naturally occurring substances such as pyrophosphate and polyphosphates, isolated from serum and urine, on the precipitation of calcium phosphates at physiological conditions has also been the subject of numerous investigations.[7-9]

1

The results of these studies suggest that these substances when present at low concentrations greatly inhibit the precipitation of calcium phosphates from supersaturated solutions. With the rather limited hydrolytic stability of the pyrophosphate ion, structurally related phosphonate compounds containing P-C-P bond in place of P-O-P bond, have been developed and their inhibitory activities on the crystal growth of calcium oxalate,[10] calcium sulfate,[11,12] barium sulfate,[13] calcium carbonate,[14-16] and calcium phosphates,[17-19] have been investigated by many workers.

The inhibitory effect of polycarboxylic acids on the precipitation of calcium phosphates has also been recently studied. Comparison of the inhibitory activity of citric acid, isocitric acid, and tricarballylic acid on calcium phosphate crystallization at physiological conditions using seeded growth technique, suggests that the hydroxyl group in the molecular backbone is a key factor in the effectiveness of these tricarboxylic acids as inhibitors.[20]

The influence of mellitic acid (benzene hexacarboxylic acid, MA) and phosphonate (hydroxyethylidene-1, 1-diphosphonic acid, EHDP) on the precipitation of calcium phosphate from solutions supersaturated with respect to all calcium phosphate phases, has also been recently investigated.[21] The results of this study indicate that although MA protects the enamel surface in acid solution and appears to have inhibiting properties similar to EHDP, EHDP is more effective as an anticalculus agent compared to MA.

Results of recent studies[18,22-25] have shown that kinetics of crystal growth of HAP as exclusive phase on different seed materials can be accurately studied by constant composition technique. Because of the recent interest in the inhibitory effect of poly-carboxylic acids on the crystallization of calcium phosphates, in the present work the constant composition method has been extended to investigate the influence of mellitic acid on the crystal growth of hydroxyapatite at low sustained supersaturation.

EXPERIMENTAL

Grade A glassware and reagent grade chemicals were used. Phosphate stock solutions were made from potassium dihydrogen phosphate and were standardized potentiometrically by titration with standard potassium hydroxide (Dil-It, J. T. Baker Co.). Calcium stock solutions were prepared from recrystallized calcium chloride dihydrate (J. T. Baker Co.) and were standardized by ion exchange method (Dowex-50W-X8) and atomic absorption spectroscopy. Mellitic acid solutions were prepared from mellitic acid (Aldrich Chemical Co.). The HAP seed crystals, prepared and characterized

as previously described,[26] were aged for at least five months
before use. The specific surface area determined by a single
point BET method using a N_2/He–30:70 gas mixture for the HAP seed
crystals was $35.8m^2g^{-1}$.

Crystal growth experiments were made in a double-walled Pyrex
cell at 37C using the constant composition technique.[27] The
stable supersaturated solutions of calcium phosphate with a molar
ratio of Tc_a: Tp=1.67, were prepared by adjusting the pH of a
premixed subsaturated solution of calcium chloride and potassium
dihydrogen phosphate to a value of 7.40 by slow addition of 0.1M
potassium hydroxide. pH measurements were made with a glass/
calomel electrode pair equilibrated at 37C. The electrode pair
was standardized before and after each experiment using NBS
standard buffer solutions.[28] The solutions were continuously
stirred (300 rpm) while nitrogen gas, presaturated with water at
37C, was bubbled through the solution to exclude carbon dioxide.
Following the inocculation with HAP seed crystals, the crystal
growth reaction was monitored by the addition of titrant solutions
from mechanically-coupled automatic buretts mounted on a modified[27]
pH-stat (Metrohm-Herisau, Model 3D combititrator, Brinkmann
Instruments, Westbury, N.Y.). The titrant solutions in the
buretts consisted of calcium chloride, potassium phosphate,
potassium hydroxide, and mellitic acid. The molar concentration
ratio of the titrant corresponded to the stoichiometry of the
HAP phase. Potassium chloride was added to the calcium phosphate
supersaturated solutions in order to maintain the ionic strength
to within 1%. The constancy of solution composition was verified
by analyzing the filtered samples which were withdrawn at various
time intervals, for calcium and phosphate according to method
described previously.[29] The rates of crystallization were
determined from the rates of addition of mixed titrants, and
corrected for surface area changes.[30]

RESULTS AND DISCUSSION

The computation of concentrations of ionic species were
made, from mass balance, electroneutrality, proton dissociation,
and equilibrium constants involving calcium ions with mellitic
acid, by iterative procedure as described previously.[31] Published
values for phosphoric acid, mellitic acid dissociation constants,[32]
calcium phosphate ion-pair association constant and calcium
mellitate equilibrium constants,[33] used in these calculations are
shown in Table I. The free energy term is computed from the
equation

$$\Delta G = - RT \ln IP/K_{so} \qquad (1)$$

in which IP is the ionic activity product and K_{so}, the value of
IP at equilibrium. R and T are the ideal gas constant and
absolute temperature. The ΔG values obtained by using the

TABLE I

CRYSTALLIZATION OF HAP ON HAP SEED CRYSTALS IN
THE PRESENCE OF MELLITIC ACID[a]

Exp #	MA x 10^{-6}M	ΔG_{DCPD}*	ΔG_{OCP}*	ΔG_{TCP}*	ΔG_{HAP}*	Rate x 10^8**
16	0.0	+4.03	+1.58	-9.94	-52.4	11.2
15	0.160	+4.03	+1.58	-9.94	-52.4	9.08
13	0.250	+4.03	+1.58	-9.94	-52.4	8.04
10	0.400	+4.03	+1.58	-9.94	-52.4	6.79
14	1.00	+4.03	+1.58	-9.93	-52.3	4.33
12	3.00	+4.04	+1.59	-9.92	-52.3	1.94

*   (kJ mole^{-1})
** (mole min^{-1}m^{-2})

Note: Initial Conditions: T_{ca}=0.500 x 10^{-3}M, Tp=0.300 x 10^{-3}M, pH=7.40, T=37C, KCl=7.00 x 10^{-3}M, volume 230 ml., Titrant=CaCl$_2$= 8.00 x 10^{-3}M; KH$_2$PO$_4$=4.80 x 10^{-3}M + KOH=9.91 x 10^{-3}M.

(H$_6$L^b=H$_5$L$^-$+H$^+$, pK=0.68; H$_5$L^{5-}=H$_4$L^{2-}+H$^+$, pK=2.21; H$_4$L^{2-}=H$_3$L^{3-}+H$^+$, pK=3.52; H$_3$L^{3-}=H$_2$L^{4-}+H$^+$, pK=5.09; H$_2$L^{4-}=HL^{5-}+H$^+$, pK=6.32; HL^{5-}= L^{6-}+H$^+$, pK=7.49 (Ref. 32); Ca^{2+}+L^{6-}=CaL^{4-}, pK=3.48; 2Ca^{2+} + L^{6-}= Ca$_2$L^{2-}, pK=2.56 (Ref. 33).

a=H$_3$PO$_4$ dissociation constants and calcium phosphate ion-pair association constant not included; see Ref. (31).

above equation for various calcium phosphate phases indicate the thermododynamic stability of the experimental solutions compared to solutions in thermodynamic equilibrium with that particular phase. Positive ΔG values represent solutions under-saturated and negative ΔG values represent solution supersaturated with respect to solid phase under consideration.

The initial conditions used in the crystal growth experiments of HAP in the presence of mellitic acid are summarized in Table I. The analytical data for a typical crystal growth experiment in the presence of mellitic acid (4.0x10^{-7}M) are shown in Table II. It can be seen that the stoichiometry of the precipitating phase was

TABLE II

CRYSTALLIZATION OF HAP ON HAP SEED CRYSTALS
AT CONSTANT SUPERSATURATION, pH=7.40, T=37C,
MELLITIC ACID=4.00x10^{-7}M, EXP 10

Time (min.)	T_{ca} 10^{-3}M	T_p 10^{-3}M	Extent of Crystallization (as % of original seed)
0	0.500	0.300	0
60	0.503	0.302	6.3
90	0.498	0.293	11.5
130	0.495	0.302	16.1
170	0.508	0.301	21.1
220	0.502	0.305	30.2

constant with calcium/phosphate molar ratio of 1.66 $\pm$ .02 for more than 3 hours of reaction. The amount of newly grown HAP phase was more than 30% of the original seed material. Plots of moles HAP/m^2 grown as a function of time for crystal growth experiments in the presence of mellitic acid, after making correction of the raw data for the observed changes in specific surface area, are shown in Figure 1. The striking constant rate of crystallization shown in Figure 1 suggests the development of a constant number of active growth sites on HAP seed crystals.

The results summarized in Table III indicate that the striking inhibitory effect of MA at low concentrations on the rate of HAP crystal growth cannot be attributed simply to the change in supersaturation due to complex formation. The curvature observed in Figure 1 at higher concentration of MA may be due to slow equilibration of the adsorbate with the surface of HAP seed crystals.

If the inhibition of HAP crystallization by mellitic acid is due to surface adsorption at active growth sites, some form of adsorption isotherm should be applicable. In many instances the Langmuir adsorption model which was developed for the adsorption of ideal gases onto solid surfaces, has been used to describe, empirically, the reduction in crystal growth rates of many sparingly soluble salts for a variety of inhibitors.

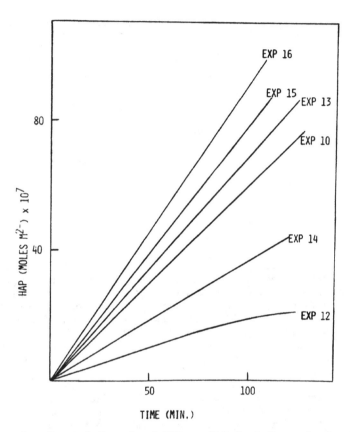

Figure 1: Crystal Growth of HAP on HAP Seed Crystals in the
           Presence of Mellitic Acid. Plots of HAP (moles m^{2-})
           as a Function of Time.

TABLE III

CRYSTALLIZATION OF HAP ON HAP SEED CRYSTALS
IN THE PRESENCE OF MELLITIC ACID

Exp	Mellitic Acid x 10^6 M	Rate x 10^8 (mole $min^{-1}m^{-2}$)	$\dfrac{R_o}{R_o-R_i}$
16	0.0	11.2	----
15	0.160	9.08	5.16
13	0.250	8.04	3.55
10	0.400	6.79	2.54
14	1.00	4.33	1.63
12	3.00	1.94	1.21

$T_{ca} = 0.500$ x 10^{-3} M; $T_p = 0.300$ x 10^3 M; pH = 7.40; T = 37C

The familiar expression describing the Langmuir adsorption of solute is given by:

$$\theta = \frac{b\,[C]}{1+b\,[C]} \qquad (2)$$

where θ is the fraction of the surface covered by adsorbing solute (in this case mellitic acid), $[C]$ is the solution concentration of the adsorbing substance, and b is the "affinity constant" for the solute-surface interaction. The rate of HAP crystallization in the presence of the mellitic acid, R_i, is proportional to the fraction of surface free from adsorbed meelitic acid, $(1-\theta)$, and is given by

$$R_i = (1-\theta)\,R_o \qquad (3)$$

where R_o is the HAP crystallization rate in the absence of mellitic acid. Equation (3) can be rearranged to give

$$\theta = \frac{R_o-R_i}{R_o} \qquad (4)$$

which on substitution in Equation (2) and upon further rearrangement, yield Equation (5)

$$\frac{R_O}{R_O - R_i} = \frac{1}{b[C]} + 1 \qquad (5)$$

According to Langmuir adsorption model, plot of R_O/R_O-R_i against $1/C$ should give a straight line. Plots of R_O/R_O-R_i against $1/C$ are shown in Figure 2 for the HAP crystal growth in the presence of MA. The excellent linearity suggests that the inhibitory effect of MA is due to adsorption at active growth sites. The value of "affinity constant" as calculated from Figure 2, is 1600×10^4 which can be compared with 98×10^4, and 24×10^4 for EHDP and pyrophosphate,[34] respectively. It is of interest to note that MA has also been found to exhibit greater inhibitory effect than EHDP on the crystal growth of fluorapatite as studied by constant composition method.[34]

Frances et al[21] studied the spontaneous precipitation of calcium phosphate ($4.0mM$ $CaCl_2$, $4.0mM$ KH_2PO_4, pH 7.40, 37C) in the presence of MA and EHDP, and concluded that EHDP is a more effective crystal growth inhibitor than MA. This is in contrast with the results obtained in the present study where only HAP exclusively grew on the HAP seed crystals in the presence of MA, compared to the formation of calcium phosphate precursor phases under spontaneous precipitation.[21] Thus it is likely that mellitic acid may have a different affect on the formation and subsequent transformation of the precursor phase than on the formation of the thermodynamically stable phase. The inhibitory effect of mellitic acid on the crystallization of CaF_2 has also been recently investigated[35] using spontaneous and constant composition techniques. The results of this study indicate that mellitic acid has only a small effect on the crystal growth of CaF_2 at high supersaturation, whereas at low supersaturation it has a striking inhibitory effect. The importance of the influence of supersaturation on the performance of inhibitors is thus clearly demonstrated in these studies.[21,34,35] Whereas the use of spontaneous precipitation method fails to demonstrate the effectiveness of inhibitors at high supersaturation, the use of constant composition method not only yields highly reproducible results but also offers the opportunity to assess the performance of inhibitors under conditions of practical importance.

In the present work it has been shown that the presence of mellitic acid at low concentration can significantly inhibit the crystal growth of HAP from calcium phosphate solutions of low supersaturation. The marked inhibitory influence of mellitic acid on the crystallization of HAP may be explained by adsorption of inhibitor molecules at the active growth sites.

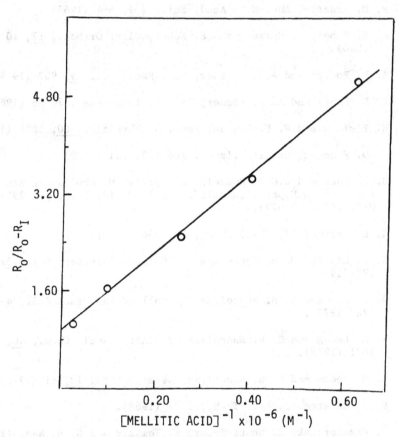

Figure 2: Crystal Growth of HAP on HAP Seed Crystals in the Presence of Mellitic Acid. Plots of R_o/R_o-R_i against $1/[Mellitic Acid]$.

REFERENCES

1. W. E. Brown, Clin. Orthop. Relat. Res., 44, 205 (1966).

2. E. D. Eanes, I. H. Gillessen, and A. S. Posner, Proc. Int.
    Conf. Crystal Growth, 373 (1966).

3. M. D. Francis, Ann. N.Y. Acad. Sci., 131, 694 (1965).

4. A. E. Sobel, M. Burger, and S. Nobel, Clin. Orthop., 17, 103
    (1960).

5. A. L. Boskey and A. S. Posner, Mat. Res. Bull., 9, 907 (1974).

6. E. D. Eanes and A. S. Posner, Calcif. Tiss. Res., 2, 38 (1968).

7. H. Fleisch and W. F. Newman, Amer. J. Physiol., 200, 1296 (1961).

8. M. D. Francis, Calcif. Tissue. Res., 3, 151 (1969).

9. H. Fleisch and R.G.G. Russel, in "Calcium Metabolism in Renal
    Failure and Nephrolithiasis", (D. S. David, Ed.), pp. 293-296,
    Wiley, N.Y. (1977).

10. G. L. Gardner, J. Phys. Chem., 82, 864 (1978).

11. S. T. Liu and G. H. Nancollas, J. Colloid Interface Sci., 52,
    593 (1975).

12. S. T. Liu and G. H. Nancollas, J. Colloid Interface Sci., 44,
    422 (1973).

13. W. H. Leung and G. H. Nancollas, J. Inorg. Nucl. Chem., 40,
    1871 (1978).

14. M. M. Reddy and G. H. Nancollas, Desalination, 12, 61 (1973).

15. P. H. Ralston, J. Pet. Tech., 1029 (1969).

16. T. Kazmierczak, E. Shuttringer, B. Tomazic and G. H. Nancollas,
    Croatica Chemica Acta, 54, 277 (1981).

17. J. P. Barone and G. H. Nancollas, J. Dent. Res., 57, 735 (1978).

18. P. G. Koutsoukos, Z. Amjad and G. H. Nancollas, J. Colloid
    Interface Sci., 83, 599 (1981).

19. J. L. Meyer and G. H. Nancollas, Calc. Tiss. Res., 13, 295
    (1973).

20. G. H. Nancollas, M. B. Tomson, G. Battaglra, H. Wawrousek and
    M. Zuckerman, in Chemistry of Wastewater Technology, (A. Rubin
    Ed.) 17, (1978), Ann Arbor Press.

21. M. D. Francis, C. L. Slough, W. W. Briner, and R. P. Oertel,
    Calcif. Tiss. Res., 23, 53 (1977).

22. P. G. Koutsoukos, Z. Amjad, M. B. Tomson, and G. H. Nancollas,
    J. Amer. Chem. Soc., 102, 1553 (1980).

23. Z. Amjad, P. G. Koutsoukos, and G. H. Nancollas, J. Dent. Res.,
    60, 1783, 1981.

24. Z. Amjad, P. G. Koutsoukos, M. B. Tomson, and G. H. Nancollas,
    J. Dent. Res., 57, 909 (1978).

25. Z. Amjad, P. G. Koutsoukos, and G. H. Nancollas, J. Dent. Res.,
    61, 1094, 1982.

26. G. H. Nancollas and M. S. Mohan, Arch. Oral Biol., 15, 731 (1970).

27. M. B. Tomson and G. H. Nancollas, Science, 200, 1059 (1978).

28. R. G. Bates, pH Determination, 2nd Ed. N.Y., John Wiley & Sons,
    1973.

29. M. B. Tomson, J. P. Barone, and G. H. Nancollas, At. Absorpt.
    Newsl., 16, 117 (1977).

30. G. H. Nancollas and P. Koutsoukos, Progress in Crystal Growth
    and Characterization, Oxford: Pergamon Press, 3, 77, 1980.

31. G. H. Nancollas, Z. Amjad, and P. Koutsoukos, Calcium Phosphates:
    Speciation, Solubility, and Kinetic Considerations, Chemical
    Modelling in Aqueous Systems, ACS Symposium Series No. 93,
    Washington, D.C., American Chemical Society, 675 (1979).

32. N. Purdie, M. B. Tomson and N. Riemann, J. Soln. Chem., 1,
    465 (1972).

33. V. A. Uchtman and R. J. Jandacek, Inorg. Chem., 19, 350 (1980).

34. Z. Amjad, unpublished data.

35. Z. Amjad, unpublished data.

ROLES OF OCTACALCIUM PHOSPHATE IN SURFACE CHEMISTRY OF APATITES

W. E. Brown, M. Mathew and L. C. Chow

American Dental Association Health Foundation Research
Unit, National Bureau of Standards, Washington, DC

INTRODUCTION

This paper reviews the effects of octacalcium phosphate (OCP), $Ca_8H_2(PO_4)_6 \cdot 5H_2O$, on the interfacial and colloidal properties of apatitic precipitates. The structural deductions are based on a combination of well established crystallographic concepts[1,2,3] and plausible projections regarding the chemical behavior of OCP.[4,5] Although the colloidal nature of the systems makes difficult the verification of these properties, the ideas provide a substantive basis for interpretation of many experimental results. Apatitic systems are of such vital importance in so many areas, and the relationships between OCP and hydroxyapatite (OHAp), $Ca_5(PO_4)_3OH$, are so close and so ubiquitous that the possibilities described here cannot be ignored. For example, the morphology of the apatitic crystallites in bones and teeth appear to derive from OCP; the consequent effects of crystallite morphology on the mechanical properties of these tissues are of great physiological importance. OCP seems to play important roles, also, in establishing the composition, solubility, reactivity, interfacial energy, nucleation, growth, and crystal-growth poisoning of apatitic materials. These all affect the surface and colloidal properties of apatitic precipitates.

STRUCTURAL CONSIDERATIONS

OCP, an acidic salt, and OHAp, a basic salt, share remarkably similar structural features despite their marked overall differences in composition, symmetry and morphology. The structures of OHAp and OCP are shown in Figures 1 and 2, respectively. The

13

published structure of OCP[6] is inexact by today's standards, but a recent refinement[7] reveals that it is esentially correct. It can be seen that the portion of the structure of OCP shown with shaded atoms corresponds very closely to a region within OHAp. Their structures have been described in detail elsewhere[4, 6, 8] and are not discussed here. Instead, we are concerned primarily with consequences arising from the similarities in the two structures.

Epitaxy

The structure of the apatitic layer of OCP (shaded area in Fig. 2) is so similar to that of OHAp that the two can grow together to form interlayered mixtures without a concomitant large increase in interfacial energy.[5,9] These interlayered mixtures give apatitic X-ray powder diffraction patterns, especially when the characteristic d_{100} peak of OCP is blocked by the X-ray beam stop, thereby frequently causing the investigator to conclude that inter and intra-crystalline mixtures containing OCP

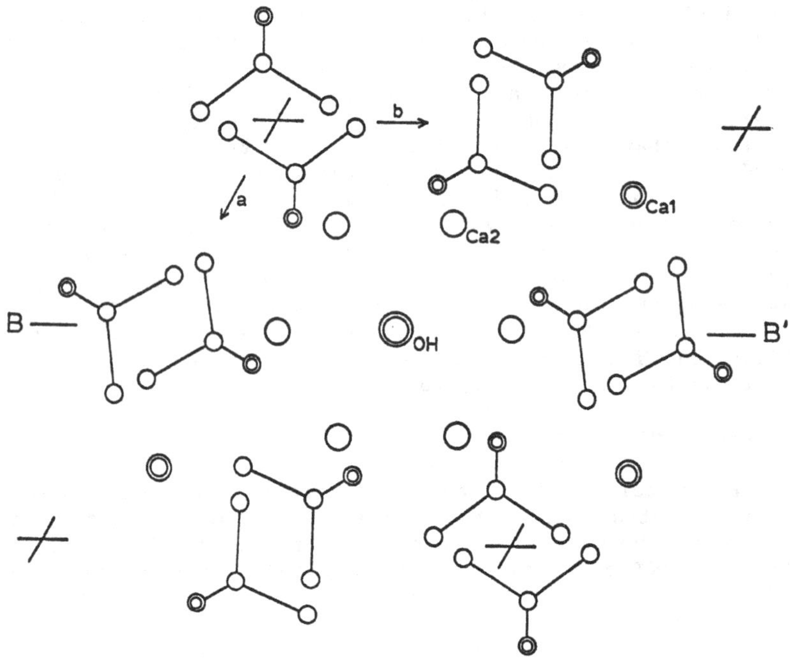

Fig. 1. The crystal structure of OHAp projected down the $\underline{c}$ axis.

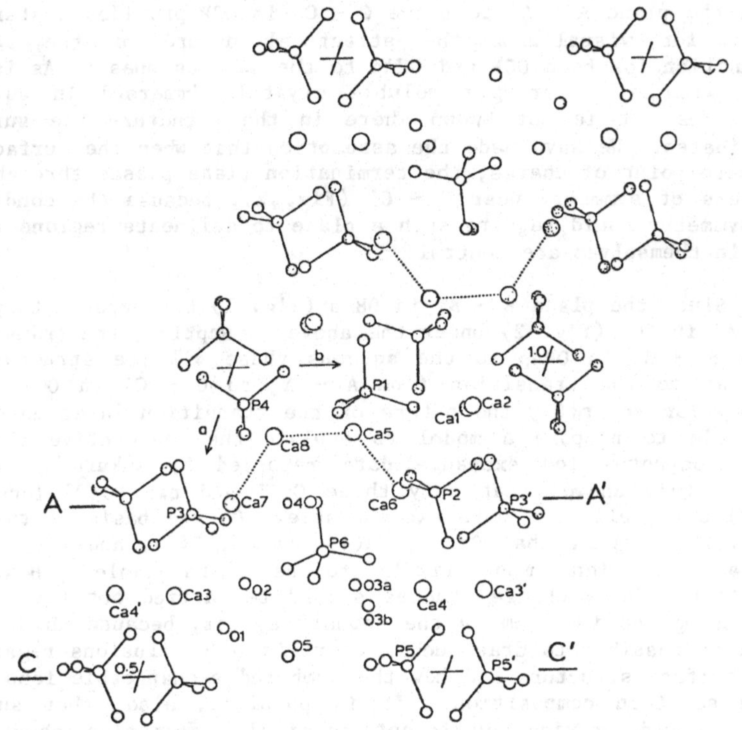

Fig. 2. The crystal structure of OCP projected down the $\underline{c}$ axis. The region with shaded atoms is very similar to that of OHAp and is referred to as the "apatitic" layer.

are exclusively nonstoichiometric hydroxyapatites. Actually, interlayered mixtures, depending on the thickness of the layers, can give two kinds of X-ray diffraction effects: (1) when layers are relatively thick, the two kinds of layers diffract independently, giving the appearance of a physical mixture of the two salts;[5] and (2) when the OCP and OHAp layers are very thin and randomly variable in thickness, the d_{h00} peaks of OCP interact with those of OHAp causing the positions of the combined peaks to shift with the Ca/P ratio of the interlayered crystals.[9]

Surface Structure

Another feature that is related to this structural similarity and to epitaxy and is of importance in the present context is the likelihood that the transition in OCP to the aqueous phase will be the same as that in OHAp. The structure from the plane A - A' to plane C - C' in OCP provides a starting point for visualizing the structural nature of the actual transition for both OCP and OHAp to the aqueous phase. As is the case with all sparingly soluble crystals immersed in aqueous solutions, it is not known where in the structure the surface terminates. We have made the assumption that when the surface is at zero point of charge, the termination plane passes through the centers of symmetry near C - C' (Fig. 2), because the condition of symmetry would require such a plane to delineate regions which within themselves are neutral.

Since the plane B - B' in OHAp (Fig. 1) corresponds to plane A - A' in OCP (Fig. 2) under the above assumption, the transition from B - B' in OHAp to the aqueous phase will be structurally similar to the transition from A - A' to C - C' in OCP. The assumption regarding the nature of the transition layer makes it possible to propose a model to explain the radioactive calcium and phosphorus ion exchange data reported by Kukura et al.[10] Their study showed that only three Ca^{2+} and two PO_4^{3-} ions per (100) unit cell face were exchangeable. On the basis of the OCP model, we suggest that Ca(3), Ca(4), Ca(6), P(5), and P(6) (Fig. 2) are the ions most likely to be exchangeable. However, additional ion-exchange studies should be carried out for points all along the isotherm in the ternary system, because this might make it possible to draw much more reliable conclusions regarding the surface structure and how the numbered exchangeable ions vary with solution composition. It is possible, also, that such a study would provide sufficient detailed information about the structure at the interface so that polymeric adhesives could be designed specifically for this surface.

Interfacial Energy

Since OHAp is a fairly hard crystal (Moh hardness = 5), one might expect it to have a moderately high interfacial energy. A good indication, however, that OHAp has an unusually low interfacial energy is the common observation that in biological minerals and in geological formations OHAp tends to form as very small crystallites which then can persist for long periods of time; in geological apatites, these periods may be millions of years. It seems highly unlikely that this could happen if OHAp had even a moderately high interfacial energy. The same considerations apply to the extremely thin ($\sim$20 Å) enamel ribbons that form as the initial precipitate in the enamel organ. We have proposed that a hydrated layer such as that from A - A' to C - C' in OCP (Fig. 2) would be especially compatible with the aqueous phase because of its hydrogen bonded structure. Such a transition layer in OHAp should reduce its interfacial energy, making it possible to persist for long periods of time in a finely divided state.

Crystallite Morphology

An important feature of the morphology of OCP is that it almost invariably occurs as some kind of platy, (100) crystals, thereby determining the nature of the dominant faces and the principal interfacial properties. In those cases where OCP acts as a precursor for the formation of OHAp, it can be said to be the genesis for the common platy, acicular, and fibrous morphologies of OHAp in which the (100)-type faces are dominant. Furthermore, as noted earlier, the transition structure in these faces probably is the same as that in the (100) face of OCP. As a result OCP can play crucial roles in establishing the morphology, growth mechanism and interfacial properties of OHAp.

The platy nature of OCP expresses itself variously in the form of platelets, elongated blades, ribbons, flimsy sheets or broad plates, frequently clustered in the form of rosettes or fascicles comprising blades or fibrous crystals.

There are three interrelated considerations (structural, kinetic, and thermodynamic) which could cause OCP to have a platy morphology. All are plausible, and all are probably involved to some degree: (1) It is a common observation, when one of the unit-cell dimensions is much larger than the others, and especially when the crystal has a pronounced layer-type structure, that the crystal tends to be platy in the plane intersected by the long axis and parallel to the layers. This may be a kinetic manifestation related to difficulty in forming a growth nucleus which, because of its thickness, requires a large number of ions to assemble during its formation. (2) The most plausible growth mechanism for OCP requires the formation of a one unit-cell thick pill box nucleus on a (100) face. Formation of such a nucleus is likely to be a relatively infrequent event, partly because of the thickness of the nucleus and partly because the attachment of Ca and PO_4 ions to the 100 face, i.e. the hydrated layer of OCP, is likely to be relatively weak. On the other hand, attachment of these ions to the lateral and terminal faces, where calcium and phosphate ions may not be as extensively hydrated, would probably be a relatively rapid process, resulting in platy morphology. (3) As noted above, the (100) faces of OCP probably have a relatively low interfacial energy, and, in accord with Wulff's law, minimization of the total interfacial energy would tend to make these the principal faces, especially when the crystallites are still very small.

Since the principal faces of both OCP and OHAp are of the (100) type, there being two such faces on an OCP crystal and six on an OHAp crystal, OCP would have a platy morphology whereas OHAp should have hexagonal acicular morphology. Thus, the

morphology of OHAp can be related to the low interfacial energy
due to the OCP transition layer in its (100) faces.

Nucleation

The initial event in the calcification of the enamel organ
is the formation of very thin, very long ribbons.[11] Although it
has been suggested that these are two unit-cell thick ribbons of
monoclinic OHAp, the interfacial energy hypothesis is more
consistent with the view that they are OCP, about one unit-cell
thick in the a direction, about twenty unit cells wide in the b
direction, and hundreds, perhaps thousands, of unit cells long in
the c direction. The interfacial energy must be a dominant
feature in the ribbon's stability, and it would seem that if it
were not for a low interfacial energy in the (100) faces, ribbons
could neither form nor persist. Hohling has proposed[12] that
these ribbons are formed by fusing of "dot-like" nuclear
particles seen in transmission electron micrographs. It seems to
us that the reverse is true - that the dot-like particles may be
formed by dehydration and decomposition of ribbons under the
vacuum and electron bombardment. The ribbons (and possibly the
dot-like particles) are important from the standpoint of the
structure of dental enamel because they eventually grow into
fibrous crystals which make enamel into the hardest tissue in the
body, yet provide it with enough resiliency so that it can
withstand brittle fracture. Similarly, the platy nature of bone
crystallites is probably related to OCP as a precursor; the
interactions between these platy crystallites and the collagen
matrix probably have a great deal to do with the strength and
mechanical properties of bone. Thus, even though OCP may be
absent or nearly absent in mature enamel and bone crystallites
(except for the transition layer in their surfaces), it may still
play a crucial role in establishing the properties of these
tissues.

HYDROLYSIS OF OCP

Except possibly under conditions of very high pressures
(e.g., ocean depths), OCP is thermodynamically less stable than
OHAp and would not form were it not for its ability to grow
rapidly. This property of OCP was discovered when calcium and
phosphate solutions were allowed to interdiffuse, and it was
found that the volumes of the resulting OCP crystals were many
orders of magnitude larger than those of the OHAp crystals.[5]
This kinetic advantage is probably the main reason why OCP
frequently forms as the initial phase instead of OHAp. On the
other hand, interfacial energy is not likely to favor one crystal
over the other if both OCP and OHAp have (100) as the dominant
forms, because they should have similar interfacial energies.

During its formation or subsequently (i.e. after it has formed), OCP tends to hydrolyze spontaneously to a nonstoichiometric OHAp. The fact that many authors have given 8/6 as the lower limit for the Ca/P ratio for nonstoichiometric apatites (also called "defect" or "calcium deficient" apatites) is another manifestation of OCP's involvement in the formation of apatitic crystals. Due to the strong similarities in the X-ray diffraction patterns, investigators frequently have not correctly distinguished between the two salts or mixtures thereof. The nonstoichiometric apatites formed by hydrolysis of OCP are almost certainly metastable with respect to well-formed OHAp, but they can persist for long periods. The attempts to establish their structures and properties have been relatively unsuccessful even though they comprise a considerable fraction of the research that has been done on apatites. We believe that there are basically three types of nonstoichiometric apatitic materials: (i) OCP-OHAp interlayered mixtures, (ii) imperfectly hydrolyzed OCP, and (iii) OHAp formed directly without involvement of OCP as a precursor. The imperfectly hydrolyzed OCP can be divided into two prototypes: (a) crystals formed under conditions where growth of the OCP crystal is not as fast as its hydrolysis so that during its formation the product contains at all times mostly defective OHAp and very little OCP; and (b) crystals formed when growth of OCP is fast compared to hydrolysis of OCP to OHAp. In the latter case, a substantial amount of OCP may be present in the product, and its subsequent hydrolysis probably would produce a relatively highly nonstoichiometric OHAp. The products formed by hydrolysis of OCP are variable in composition and as yet poorly understood. However, as a matter of experience it appears that whenever OCP acts as a precursor to OHAp, the resulting apatite is nonstoichiometric and usually contains impurities and defects incorporated during the hydrolysis process.[13,14] In the following, we examine some of the colloidal and surface properties that may be imparted to the product OHAp during the imperfect hydrolysis process.

Defective Hydrolysis Products

There are certain conditions of preparation where OCP cannot participate as a precursor in the formation of OHAp, high temperature, high pH, presence of fluoride ions, and very low levels of supersaturation. In all these cases, the resultant apatite tends to have the stoichiometric Ca/P ratio. On the other hand, most conditions of synthesis of OHAp are so supersaturated that OCP can participate as a precursor, and in these the product OHAp tends to have low Ca/P ratios. It is of value, therefore, to examine the OCP hydrolysis reaction in more detail. The hydrolysis of OCP to stoichiometric OHAp in the presence and the absence, respectively, of calcium ions may be written as follows:

$$Ca_8H_2(PO_4)_6 \cdot 5H_2O + 2Ca^{2+} + 4OH^- = 2Ca_5(PO_4)_3OH + 7H_2O \quad (1)$$

$$5/4 \ Ca_8H_2(PO_4)_6 \cdot 5H_2O + 9/2 \ OH^- = 2Ca_5(PO_4)_3OH$$
$$+ 3/2 \ PO_4^{3-} + 35/4 \ H_2O \quad (2)$$

It can be seen from equation (2) that the hydrolysis process in the absence of calcium involves loss of phosphate ions from the crystal. Depending on the mechanism of hydrolysis (e.g., whether the phosphate ions are lost from the (100) faces or from the edges of the platy crystals), this process could profoundly alter the surface properties as compared to those that would obtain if the hydrolysis were as depicted by equation (1). This would be additional to effects that would result from formation of a non-stoichiometric OHAp. Furthermore, our experience has been that the hydrolysis predicted by equation (2) is seldom complete; in fact, it may cease after approximately 50% of the phosphate indicated by equation (2) has been released by the crystals. Magnesium ions and many other adsorbents are known to impede the hydrolysis process.[5,13] The effects of adsorbents on the kinetics of growth and hydrolysis of OCP is an area of research which deserves considerably more study because of its biological implications, especially since some of these adsorbed ions become incorporated into the product.

Structural Effects

In Figure 3a is depicted a crystallite of OCP two unit cells thick; this unit then hydrolyzes by a mechanism in which it is hypothesized that a layer one unit cell thick of OCP forms a layer two unit cells thick of OHAp (Figs. 3b and 3c). Since d_{100} of OCP is 18.64 Å and $2d_{100}$ of OHAp is 16.32 Å, the hydrolysis of a unit-cell layer of OCP to a two-unit-cell layer of OHAp would, in the ideal case, require a shrinkage of 2.36 Å in the direction of the reciprocal axis $\underline{a}^*$. This shrinkage is also accompanied by displacement of the newly formed OHAp layer relative to the adjacent OHAp layer by about 3.12 Å in the $\underline{b}$ direction (Fig. 3b). If instead of closing this gap so that the layers remain separated, as shown in Figure 3c, a crevice with a spacing of about 2.36 Å would be formed. This crevice, especially if it were enlarged by coalescence of several such crevices, could produce an internal surface of the type implied by Hendricks and Hill.[15] Furthermore, entry of ions or water molecules into these crevices could expand them in the same way that is known to occur in clays. These water molecules, as well as the structural water present in the hydrated layers of OCP adjacent to these crevices, could account in part for the high degrees of hydration of many apatitic precipitates. The crevices may also be sites in which impurities such as carbonate ions enter into these crystals, in which case they would not occupy true lattice sites. If this

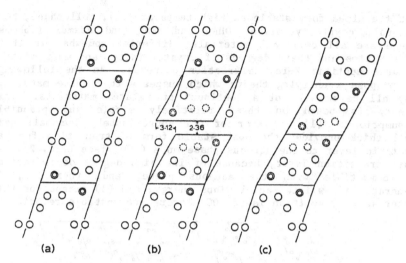

Fig. 3. Proposed scheme for hydrolysis of OCP to OHAp. Only the Ca ions are shown: (a) two unit cells of OCP; (b) formation of two unit cells of OHAp from the inner unit cell of OCP prior to layer shifts; and (c) after rearrangement of the two OHAp unit cells.

were so, the crevices may be the locations for part of the "surface" carbonate.[16] It is of interest that most of the apatites which were found by Young[17] to exchange hydroxyl ions relatively rapidly are the ones in which OCP could have acted as a precursor in their formation, indicating that crevices may have been present in these materials. The dissolution of the inner cores of enamel crystallites during caries attack may be another manifestation of the presence of a crevice at the planar site which corresponds to the initial ribbon that forms initially when enamel mineralizes.[11] Similarly, the dark line observed at this site in electron micrographs of enamel crystallites may be caused by a crevice, which in this case would be called a dislocation. It is apparent that the "internal surface" associated with those crevices could contribute in important ways to the properties of apatites.

APATITIC COLLOIDS

Over the years a number of colloidal apatitic materials have been encountered which are sufficiently unique that they have been given appelations descriptive of their properties. These include amorphous calcium phosphate (ACP),[18] tricalcium phosphate hydrate[19] (also, α-tricalcium phosphate, which is hydrated and not the alpha form stable at high temperatures), collophane, bone mineral, cryptocrystalline OHAp and OCP, and enamel ribbons. These are all colloidal materials, differing somewhat in their Ca/P ratios and their degrees of crystallinity, and most of them produce apatitic X-ray diffraction patterns. In the following, we propose a unifying theory which suggests that these materials may all be members of a continuum of related materials. This theory is based on three relatively simple and plausible assumptions: (1) A layer of OCP approximately one half unit cell thick provides the most stable form of transition from an apatitic layer to the aqueous phase on a (100) face (Fig. 2), (2) this transition layer, because of its high degree of hydration, is compatible with the aqueous phase, and, therefore, it generates a low interfacial Gibbs energy, and (3) except for this interfacial transition layer, OCP layers are unstable relative to

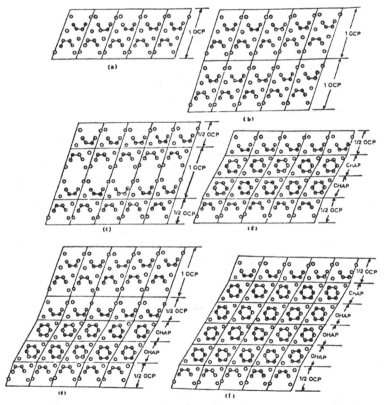

Fig. 4. Proposed mechanism for the growth of OCP-OHAp crystals in which the formation of a unit cell thick layer of OCP alternates with its hydrolysis to two unit cell thick layers of OHAp (see text).

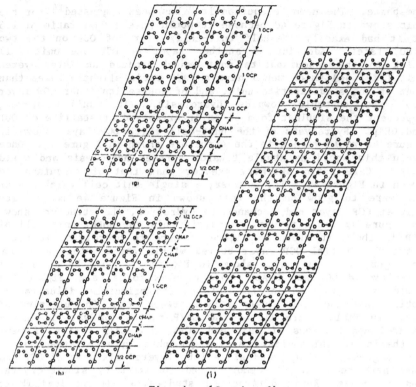

Fig. 4. (Continued).

OHAp and may hydrolyze spontaneously. In Figure 4a is depicted a
cross section of a ribbon or platelet of OCP one unit cell thick
and five unit cells wide projected down the $\underline{c}$ axis. Because this
crystallite comprises two half unit cell thicknesses of OCP with
their hydrated layers adjacent to the aqueous phase, it would be
stable towards further hydrolysis to OHAp. In Figure 4b is shown
the crystallite after further growth so that it is now two unit
cells thick.   In Figure 4c is shown the same crystallite as in
Figure 4b except that now we conceive of it as a single unit cell
thickness of OCP flanked on either side by a half unit cell
thickness of OCP.   In this representation, it can be seen that
the two outside half-cell-layers should be stable towards further
hydrolysis, but this would not be true for the inner unit-cell
layer of OCP.   If the inner layer were to hydrolyze to OHAp in
the manner shown in Figure 4d, the unit would once again be
stable toward further hydrolysis because (i) the outside of the
(100) face of the crystallite would consist of a half unit cell
of OCP and (ii) the inside now is a layer of OHAp, which under
most biological conditions would be the stable form of calcium

phosphate. The name "sesquiapatite" has been suggested[14] for the unit shown in Figure 4d because it would have a Ca/P ratio of 1.5 if it had exactly one-half unit-cell layers of OCP on the two (100) faces. Addition of another layer of OCP one unit cell thick to the sesquiapatite unit would produce an interlayered unit (Fig. 4e) which would have a Ca/P ratio slightly lower than that of the sesquiapatite unit, and of course its inner OCP layer would be unstable towards hydrolysis. It would, however, represent the simplest form of an interlayered crystallite of OCP and OHAp. Hydrolysis of the unit-cell thick OCP layer shown in Figure 4e would result in the structure shown in Figure 4f. Once again this would be stable towards further hydrolysis and would have a Ca/P ratio slightly greater than that of sesquiapatite shown in Figure 4d. If, however, a single unit cell thickness of OCP were to grow on the unit shown in Figure 4e before its initial OCP layer had a chance to hydrolyze, the structure shown in Figure 4g would result. Then, if only the outer unit cell of OCP in the unit shown in Figure 4g were to hydrolyze, it would leave the initial OCP layer buried within the crystal, as shown in Figure 4h. The units shown in Figure 4h would be a form of interlayered OCP and OHAp. More complex forms (e.g. Figure 4i) could be constructed on the basis of the principles illustrated, noting that the OHAp layers should always occur in multiples of two. In Table 1 are listed the Ca/P ratios for various inter-layered combinations of OHAp and OCP. The Ca/P ratios of many synthetic and biological calcium phosphate precipitates are in the range 1.4 to 1.67; and since interlayered mixtures of OCP and OHAp have been clearly demonstrated to exist by single-crystal and by powder X-ray diffraction studies,[5,9] it is logical to expect that low Ca/P ratios are related in part to interlayering. However, the situation is probably more complex than simple combinations of layers of idealized OCP and OHAp. The hydrolysis of OCP to OHAp is an irreversible, topotactic process, and almost

Table 1.   The Ca/P Ratio and Water Content as Function of
           Number of OHAp Unit Cells Per Unit Cell of OCP

No. of OHAp Unit Cells	Ca/P Ratio	Water Molecules/ Per 6 PO_4
0	1.33	5
2	1.5	2.5
4	1.56	1.67
6	1.58	1.25
8	1.6	1
10	1.61	0.83
20	1.64	0.45
40	1.65	0.24

certainly does not produce a defect-free OHAp. The nature and extent of these defects in the apatitic layers and interlayering with OCP are likely to have considerable influence on the chemical and surface properties of these products.

In the following we describe some natural and synthetic materials which may be interlayered and partially hydrolyzed products that can be described on the basis of the principles given above.

Amorphous calcium phosphate (ACP) tends to have a Ca/P ratio approximating 1.5, especially when prepared at high pH, but ratios as low as 1.0 have been reported.[20] Its X-ray diffraction pattern consists of a broad peak at about 30° in 2θ (CuK$_\alpha$ radiation) and its radial distribution function suggests that it has short range order up to about 9.5 Å.[18] Posner and co-workers[18] have proposed that it comprises clusters with composition $Ca_3(PO_4)_3Ca_3(PO_4)_3Ca_3$ in which the ionic configuration simulates the triangular arrangement of these ions around the hexad axis of OHAp. Thus the cluster would consist of alternating calcium and phosphate triangles. As an alternative proposal, we have suggested[4] that the ACP structure is based more on the sesquiapatite-like prototype units of the type shown in Fig. 4d because of the stabilizing effect of the OCP half layers. However, ACP is known to transform into OCP in a relatively short period of time, and one might question why this would happen if ACP had an OCP-like structure. The most plausible explanation is that in ACP the cryptocrystalline particles are very small platelets so that the areas of the faces that form the lateral and terminal edges are significant compared to those of the (100) faces. The higher interfacial energy that is probably associated with the edge and terminal faces would provide the driving force for Ostwald ripening of ACP particles into broader OCP platelets.

The synthetic precipitate "tricalcium phosphate hydrate"[19] which also tends to have a Ca/P ratio near 1.5, produces a poorly-resolved, apatitic X-ray diffraction pattern, but it is more stable chemically than ACP. Again, if the stabilizing effect of the outer one-half unit-cell of OCP is valid, hydrated tricalcium phosphate could be a mixture of the materials represented in Table 1, and the high degree of hydration would derive in part from the hydrated layer of OCP. Similarly, apatitic crystallites of bone, dentin, cementum, renal stones, calcified pineal glands, dental calculus, and other pathological deposits, which are somewhat more crystalline than hydrated tricalcium phosphate, could also be mixtures of products such as those in Table 1. In these, the number of unit cell layers of apatite per unit cell of OCP would tend to be in the range 4 to 8. An advantage of this proposition is that all these materials would be structurally related by a simple set of principles regarding interfacial energies of OCP-related materials, so that it is not necessary to generate new structural entities. Although structurally related, their Ca/P ratios could vary considerably for reasons given later.

DISCUSSION

The platy morphologies of tooth, bone, renal stones, dental calculus and many synthetic apatite crystallites strongly indicate that OCP is frequently the nucleation agent, and as a result establishes which crystallite faces will dominate the resulting crystals as well as crystal size and composition. In doing so, OCP would be the principal source for many of the interfacial, colloidal, chemical, and physical properties of the product OHAp. During the crystal growth process (i.e. after nucleation has occurred), OCP could play important roles by acting as a precursor that then hydrolyzes into an impure, defective OHAp. The hydrolysis can occur either while the crystals are growing or after they are formed. The impurities and defects incorporated may affect the level of electrical imbalance within the crystallites, which in turn could have considerable influence on the Galvani potential of that phase and thus on the conditions leading to the zero points of charge on a given face. They probably account also for much of the variability in solubility ($K_{sp} = 10^{-66}$ to 10^{-55}) of OHAp.

The remarkable ribbon-like entity that forms initially in enamel, and which probably owes its existence to the presence of the hydrated layer of OCP on its (100) faces, is an important clue regarding the nature of the colloidal calcium phosphates. It is a well-characterized prototype for an end member of the series of materials for which colloidal apatites are the other extreme. Similarly, the persistence of colloidal apatitic minerals for geological periods of time becomes more understandable if they have low interfacial energies imparted by the hydrated layer on the (100) of OCP. On the other hand, we have offered an ad hoc explanation of why ACP can be so labile chemically even if it has an OCP-like structure: the platy dimensions of the ACP particles may be so small that interfacial energies of the lateral and terminal edges may elevate the Gibbs energies of these particles.

It has been proposed[4] that four factors affect the Ca/P ratios of apatitic precipitates: (1) electrochemical equilibria cause variations in the Ca/P ratio of the outer layer (i.e. Stern layer) which might be considered to be adsorbed calcium and/or phosphate ions; (2) OCP-OHAp interlayered mixtures of the type listed in Table 1; (3) the half-unit cells of OCP at the (100) surfaces; and (4) the presence of imperfectly hydrolyzed OCP. The latter products could be in the form of sequiapatite-like materials derived from the hydrolysis of pairs of OCP unit-cell layers or some other kind of calcium-deficient apatite in which the crystal lattice would be that of OHAp, but all of the lattice sites would not be filled. From this listing it is obvious that these apatitic colloidal precipitates have several sources of potential variability in composition which could affect their properties immensely. Sophisticated techniques and interpretations will be needed to distinguish between the effects produced by the various mechanisms.

In a recent paper by Williams and Sallis,[21] it was claimed that additives affected the maturation of ACP in either of two ways which they attributed to the presence of two types of "adsorption sites." However, they disregarded one important aspect of the maturation of ACP, namely the formation of OCP as an intermediary. The maturation of ACP can be represented by the following diagram

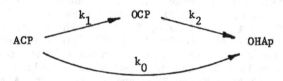

where k_1 is the rate of conversion of ACP to OCP, k_2 is the rate of hydrolysis of OCP to OHAp, and k_0 is the rate of direct conversion of ACP to OHAp. An absorbent which greatly reduces k_0 and k_2 relative to k_1 will, in effect, stop the maturation in the OCP stage. Conversely, a rapid k_2 or k_0 relative to k_1 would cause the principal product to be OHAp. This phenomenon, which involves the surface chemistry of ACP, OCP, and OHAp, may prove to be of great physiological importance, because it may provide the means, through the use of surface active chemicals, to control the nature of the calcium phosphate phases in hard tissues in advantageous ways.

References

1. J. Christoffersen and M. R. Christoffersen, Kinetics of dissolution of calcium hydroxyapatite IV. The effect of some biologically important inhibitors, J. Cryst. Growth 53:42 (1981).
2. G. H. Nancollas, Enamel apatite nucleation and crystal growth, J. Dent. Res. 58(B):861 (1979).
3. C. F. Frank, Capillary equilibria of dislocated crystals, Acta Crystallogr. 4:497 (1951).

4. W. E. Brown, M. Mathew and M. S. Tung, The crystal chemistry of octacalcium phosphate, Prog. Cryst. Growth Charact. 4:59 (1981).

5. W. E. Brown, J. R. Lehr, J. P. Smith and A. W. Frazier, Crystallographic and chemical relations between octacalcium phosphate and hydroxyapatite, Nature (London) 196:1050 (1962).

6. W. E. Brown, The crystal structure of octacalcium phosphate, Nature (London) 196:1048 (1962).

7. M. Mathew and W. E. Brown unpublished results.

8. M. I. Kay, R. A. Young and A. S. Posner, Crystal structure of hydroxyapatite, Nature (London) 204:1050 (1964).

9. W. E. Brown, L. W. Schroeder and J. S. Ferris, Interlayering of crystalline octacalcium phosphate and hydroxyapatite, J. Phys. Chem. 83:1385 (1979).

10. M. Kukura, L. C. Bell, A. M. Posner and J. P. Quirk, Kinetics of isotope exchange on hydroxyapatite, Soil Sci. Soc. Amer., Proc. 37:364 (1973).

11. M. U. Nylen, E. D. Eanes and K. A. Omnell, Crystal growth in rat enamel, J. Cell Biol. 18:109 (1963).

12. H. J. Hohling, J. Althoff, R. H. Barckhaus, E. -R. Krefting, G. Lissner and P. Quint (1981) "Early stages of crystal nucleation in hard tissue formation," International Cell Biology 1980-81, Ed. H. G. Schweiger, Berlin: Springer.

13. N. S. Chickerur, M. S. Tung and W. E. Brown, A mechanism for incorporation of carbonate into apatite, Calcif. Tissue Int. 32:55 (1980).

14. W. E. Brown, M. S. Tung and L. C. Chow, Role of octacalcium phosphate in the incorporation of impurities into apatites, International Congress on Phosphorus Compounds, Boston, 59-71 (1980).

15. S. B. Hendricks and W. L. Hill, Nature of bone and phosphate rock, Proc. Nat. Acad. Sci. (U.S.) 36:731 (1950).

16. W. F. Neuman and M. W. Neuman, "The Chemical Dynamics of Bone Mineral," Univ. Chicago Press, Chicago, 1958.

17. D. W. Holcomb and R. A. Young, Thermal decomposition of human tooth enamel, Calcif. Tissue Int. 31:189 (1980).

18. A. S. Posner and F. Betts, Synthetic amorphous calcium phosphate and its relation to bone mineral structure, Accounts Chem. Res. 9:273 (1975).

19. M. J. Dallemagne and C. Fabry, Structure of bone salts, Ciba Foundation Symposium on Bone Structure and Metabolism, 1955, 14-35 (1956).

20. M. D. Francis and N. C. Webb, Hydroxyapatite formation from hydrated calcium monohydrogen phosphate precursor, Calcif. Tissue Res. 6:335 (1971).

21. G. Williams and J. D. Sallis, Structural factors influencing the ability of compounds to inhibit hydroxyapatite formation, Calcif. Tissue Int. 34:169 (1982).

SOLUBILITY AND INTERFACIAL PROPERTIES OF HYDROXYAPATITE: A REVIEW

S. Chander and D. W. Fuerstenau

Department of Materials Science and Mineral Engineering
University of California, Berkeley, CA 94720

ABSTRACT

Recent literature on solubility and interfacial properties of
hydroxyapatite is reviewed. Representation of thermodynamic equilibrium between the solid and the solution is discussed for the
system containing calcium, phosphate and hydrogen ions and their
complexes. The origin of surface charge and the conditions for
zero charge on hydroxyapatite are discussed, and adsorption mechanisms for fluoride and organic ions are critically reviewed. The
literature review reveals that complex interfacial and dissolution
interactions occur in this system.

INTRODUCTION

The complex nature of calcium phosphates is evident in the
confusion and contradictions that have existed in the large amount
of research reported by numerous investigators over more than two
centuries. Development in the understanding of the character and
properties of these materials has been hampered by several factors
which include: the co-existence of many bulk and surface phases,
the possibility of various phases being present in solid solutions,
occurrence of non-stoichiometric phases, lattice substitutions by
the impurities, the existence of several simple and complex ions in
solution, among others. Nonequilibrium conditions, that is the
slow attainment of equilibrium in these systems, and the role of
unaccounted impurities have also contributed to the slow development in understanding the properties of calcium phosphates. The
complex character of these materials is also evident in that investigations of their behavior have closely followed theoretical and

instrumental developments in the general field of chemistry. A historical development of the properties of calcium phosphates will not be attempted here. The objective of this review is to discuss the more recent literature on the solubility and interfacial properties. These properties are important in understanding the behavior of calcium phosphates in such diverse areas as biology, geology, mineral processing, soil science and waste treatment processes.

THERMODYNAMIC SOLUBILITY

Values for the solubility product of hydroxyapatite (HA) obtained by various investigators are listed in Table 1. Although controversies have existed in the literature, as reviewed by several researchers (1-3), it is now fairly well established that stoichiometrically pure hydroxyapatite with well-defined crystal structure has a thermodynamic solubility product, the most probable value being $pK_s = 115$. Impurities, particularly carbonate (26), have contributed to discrepancies in the observed results when equilibrium is approached in undersaturated solutions. When studies are made by precipitation from supersaturated solutions, equilibrium is not achieved unless long times or elevated temperatures are used.

Several precursor phases have been identified by investigators studying the precipitation of HA. These phases include amorphous calcium phosphate (ACP) (27), tricalcium phosphate (TCP) (28-29), octacalcium phosphate (OCP) (30-32), and dicalcium phosphate dihydrate (DCPD) (33). Recent studies show that hydroxyapatite can be directly precipitated from solutions of low supersaturation (32,34). Only at high supersaturation are intermediates formed. Van den Hoek et al. (35) explained the kinetics of precipitation of fluorapatite in terms of a model which assumes that the deposition of successive layers onto the precipitate surface occurs with each layer in equilibrium with the bulk solution at the time of deposition. Their results indicate that equilibrium in the solid phase does not exist, and therefore non-stoichiometric apatites with defect structure are formed. Studies of Amjad et al. (36) and Chander et al. (37) suggest that the growth of fluorapatite is proportional to the degree of supersaturation. These results show that the growth kinetics for fluorapatite under conditions of low supersaturation may be explained without invoking the formation of a precursor phase. Apparently, the precursor phase is important only in the nucleation stage.

SOLID/AQUEOUS SOLUTION EQUILIBRIUM

When hydroxyapatite and other calcium phosphates dissolve, the lattice ions may undergo various hydrolysis and complexation reac-

Table 1. Published Determinations of Solubility
Product of Hydroxyapatite.

Investigator (year)	pKs[**]	Temp.°C	Remarks	Ref.
Holt et al. (1925)	111	38	A	4
Bjerrum (1949)	115-117		--	6
Elmore et al. (1950)	112	25	--	7
Clark (1955)	115.5	25	B,C	8
Levinskas and Neuman (1955)	115	24	D	9
Hayek et al. (1958)	89,81	18,40	--	10
Brown (1960)	111		C	11
Lindsay and Moreno(1960)	114	25	C	12
Olsen et al.(1960)	111	25	C	13
Rootare et al. (1962)	108-122	40	E	5
Hagen (1965)	114	37	--	14
Fassbender et al. (1966)	113	25	--	15
Moreno et al. (1968)	115	25	C	16
Weir et al. (1971)	117	25	--	17
Chien (1972)	121	25	--	18
Saleeb and deBruyn(1972)	115		C	19
Avnimelech et al. (1973)	116	25	C	20
Smith et al. (1976)	117	20	C	21
Wu et al. (1976)	125	30	F	22
Brown et al. (1977)	114		--	23
McDowell et al.	177	25	C	24
Fawzi et al.(1978)	122	30	F	25

$$**pK_s = -\log \left([Ca]^{10}[PO_4]^6[OH]^2 \right)$$

A: Calculated by Rootare et al.(5) D: Calculated by Kibby and Hall(3)
B: Precipitation   E: Function of Amount Dissolved
C: Dissolution Equilibrium   F: Dissolution Kinetics

tions in solution. A list of such reactions is given in Table 2
along with corresponding equilibrium relations between the concentration of the various solution species. The adsorption of simple
and complex ions on the surface of the solid determines the charge
characteristics of the solid/solution interface as well as the
amount of solid dissolved under a given set of solution conditions. Graphical representation of solid/solution equilibria is a
convenient way to represent various chemical reactions between a
solid and the solution. For a system of three or more independent
variables, the selection of independent variables to draw two-
dimensional diagrams is, to some extent, arbitrary. A number of
diagrams were drawn by Chander and Fuerstenau with different choice
of independent variables and the merits of each were discussed (38).

In interfacial phenomena, charged species, especially the ions
common to the solid and the solution and their complexes, determine
the charge characteristics of the solid/solution interface. Accor-
dingly, diagrams drawn with ionic species as the independent
variables are desirable. The equilibrium diagram based on pH and
the activities of the predominant calcium and phosphorous species
in solution as the independent variables is given in Figure 1,
which is a modification of a diagram in Reference 37. This diagram
is a two-dimensional projection of the solubility surface with the
pH axis being perpendicular to the plane of the paper.   The
parallel lines are the pH-invariant lines drawn at an interval of
0.1 pH unit. The solubility surface consists of planes marked by
letters A to I. The plane I shows the increase in solubility of HA
with increasing pH.   The chemical reaction and the equilibrium
relation for each plane are listed in Table 2. The planes in this
figure intersect to give a line which is referred to as a
<u>solubility edge</u>. In reality, the transition from one plane to the
other will be more gradual. Each solubility edge corresponds to a
change in the solution species. For example, solubility edge A/B
is for the equilibrium, $H_2PO_4^-/HPO_4^=$. The ions in equilibrium at
the various solubility edges are listed in Table 3.   Other
predominant solution ions are also listed in this table. The plane
M marks the transformation of hydroxyapatite to monenite. This
plane is perpendicular to the pH axis and is located at pH 4.6.

ORIGIN OF SURFACE CHARGE

The charge of ionic solids is considered to be the result of
preferential dissolution or adorption of lattice ions, which are
called the potential-determining ions. The species which form
complexes with the lattice ions may also contribute to the surface
charge. Complex ions containing lattice ions may either be
directly produced at the solid/solution interface or may form in
solution and subsequently adsorb on the solid surface in amounts
proportional to their concentrations in solution. In either case,

Table 2. List of Chemical Reactions and Equilibrium Relations in the Hydroxyapatite-Aqueous Solution System

Solubility Plane	Chemical Reaction and Equilibrium Relation
A	$Ca_{10}(PO_4)_6(OH)_2 + 14\ H^+ = Ca^{++} + 6\ H_2PO_4^- + 2\ H_2O$ $pCa + 0.6\ p(H_2PO_4) - 1.4\ pH = -2.66$
B	$Ca_{10}(PO_4)_6(OH)_2 + 8\ H^+ = 10\ Ca^{++} + 6\ HPO_4^= + 2\ H_2O$ $pCa + 0.6\ p(HPO_4) - 0.8\ pH = 1.54$
C	$Ca_{10}(PO_4)_6(OH)_2 + 14\ H^+ = 6\ CaH_2PO_4^+ + 4\ Ca^{++} + 2\ H_2O$ $p(CaH_2PO_4) + 0.667\ pCa - 2.333\ pH = -5.51$
D	$Ca_{10}(PO_4)_6(OH)_2 + 4\ H_2PO_4^- + 14\ H^+ = 10\ CaH_2PO_4^+ + 2\ H_2O$ $p(CaH_2PO_4) - 1.4\ pH - 0.4\ p(H_2PO_4) = -3.74$
E	$Ca_{10}(PO_4)_6(OH)_2 + 4\ H_2PO_4^- + 4\ H^+ = 10\ CaHPO_4(aq) + 2H_2O$ $p(CaHPO_4) - 0.4\ p(H_2PO_4) - 0.4\ pH = 1.64$
F	$Ca_{10}(PO_4)_6(OH)_2 + 8\ H^+ = 6\ CaHPO_4(aq) + 4\ Ca^{++} + 2\ H_2O$ $p(CaHPO_4) + 0.667\ pCa - 1.333\ pH = -0.133$
G	$Ca_{10}(PO_4)_6(OH)_2 + 8\ H^+ + 4\ HPO_4^= = 10\ CaHPO_4(aq) + 2\ H_2O$ $p(CaHPO_4) - 0.8\ pH - 0.4\ p(HPO_4) = -1.16$
H	$Ca_{10}(PO_4)_6(OH)_2 + 2\ H^+ = 4\ Ca^{++} + 6\ CaPO_4^- + 2\ H_2O$ $p(CaPO_4) + 0.667\ pCa - 0.333\ pH = 8.107$
I	$Ca_{10}(PO_4)_6(OH)_2 + 4\ HPO_4^= = 10\ CaPO_4^- + 2\ H^+ + 2\ H_2O$ $p(CaPO_4) + 0.2\ pH - 0.4\ p(HPO_4) = 7.08$

the net result will be an alteration of the surface charge in direct relation to the charge on the complex ions.

The charge at the surface of a solid containing a lattice cation and a lattice anion of the same valence is zero when the surface concentrations (or adsorption densities) of the two ions are equal (when valencies are different, the equivalent surface charge contribution from the adsorbed lattice cations must be equal to the contribution from the anions). The solution conditions where this situation occurs is referred to as the point of zero charge (PZC). In dealing with solids containing three lattice ions, the condition of zero charge may be obtained by changing the

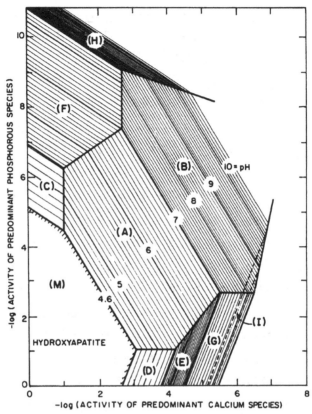

Fig. 1.   Two-dimensional projection of the solubility surface of
           hydroxyapatite consisting of planes (A) through (I). The
           parallel lines on the solubility surface are the pH-
           invariant lines drawn at an interval of 0.1 pH unit.
           Below pH 4.6 hydroxyapatite transforms to monenite; this
           transformation is shown by plane (M). The predominant
           calcium and phosphorous species in equilibrium with
           hydroxyapatite along various planes are: $A(Ca^{++}, H_2PO_4^{-})$;
           $B(Ca^{++}, HPO_4^{=})$; $C(Ca^{++}, CaH_2PO_4^{+})$; $D(CaH_2PO_4^{+}, H_2PO_4^{-})$;
           $E(CaHPO_4, H_2PO_4^{-})$; $F(Ca^{++}, CaHPO_4)$; $G(CaHPO_4, HPO_4^{=})$;
           $H(Ca^{++}, CaPO_4^{-})$ and $I(CaPO_4^{-}, HPO_4^{=})$.

Table 3. List of Species in Equilibrium with Hydroxyapatite at
Various Solubility Edges

Solubility Edge	Ions in Equilibrium at the Solubility Edge	Other Predominant Solution Ions
A/B	$H_2PO_4^-/HPO_4^=$	Ca^{++}
A/C	$H_2PO_4^-/CaH_2PO_4^+$	Ca^{++}, H^+
A/D	$Ca^{++}/CaH_2PO_4^+$	$H_2PO_4^-$, H^+
A/E	$Ca^{++}/CaHPO_4$	$H_2PO_4^-$, H^+
A/F	$H_2PO_4^-/CaHPO_4$	Ca^{++}, H^+
B/G	$Ca^{++}/CaHPO_4$	HPO_4^-, OH^-
B/H	$HPO_4^=/CaPO_4^-$	Ca^{++}, OH^-
B/I	$Ca^{++}/CaPO_4^-$	$HPO_4^=$, OH^-
C/F	$CaH_2PO_4^+/CaHPO_4$	Ca^{++}, H^+
D/E	$CaH_2PO_4^+/CaHPO_4$	$H_2PO_4^-$, H^+
E/G	$H_2PO_4^-/HPO_4^=$	$CaHPO_4$
F/H	$CaHPO_4/CaPO_4^-$	Ca^{++}, OH^-
G/I	$CaHPO_4/CaPO_4^-$	$HPO_4^=$, OH^-

concentration of two of the ions independently. This leads to a
set of points where the surface charge is zero, giving rise to a
line of zero charge on the solubility surface (19). Characterization of the electrical double layer properties of apatites,
therefore, requires determination of the line of zero charge (LZC).

Attempts to calculate PZC's from thermodynamic properties have
been partially successful. Parks carried out such calculations for
oxides (39) and complex silicate minerals (40). For a number of
oxidic solids the assumption that all hydroxy complexes adsorb onto
the surface with about equal equilibrium constants gives fairly
good agreement between experimental and calculated PZC's. For many
ionic solids, such as AgI, AgBr, AgCl, CaF_2, etc., the experimental
point of zero charge differs considerably from the equivalence
point (41), suggesting higher affinity of one or the other ion.
Roman et al. (42) attributed the difference in affinity of ions for
the surface to differences in their hydration characteristics.

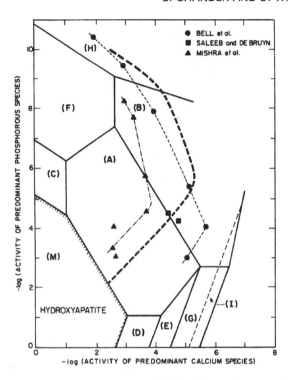

Fig. 2. The position of the calculated line of zero charge (shown
by the heavy dashed line) and the experimental points of
zero charge on the solubility surface of hydroxyapatite.
The planes (A) to (I) and (M) are the same as in Figure 1.

Miller and Hiskey (43) suggested that the potential-determining
anions and cations may have different energies when they are at the
surface, which should also be taken into consideration.

The line of zero charge for hydroxyapatite was calculated with
the assumption that the affinity for the surface of all the lattice
ions in solution can be considered essentially equal. The line of
zero charge is shown as a heavy line on the solubility surface in
Figure 2. The experimental PZC data points of Saleeb and deBruyn
(19), Bell et al. (44) and Mishra et al. (45) are also plotted in
the figure. In plotting the experimental data points, it has been
assumed that the solution is in thermodynamic equilibrium with the
solid phase. The heavy broken lines through the data of Bell et al.
and Mishra et al. may be referred to as the experimental LZC's.
Although the calculated and experimental LZC's do not match,
similarities between the two are pronounced. The general shape of
the two LZC's is the same. Bell's LZC and the calculated LZC
intersect at about pH 8.5, the PZC obtained without addition of any

calcium or phosphorous containing salt to the system. It is tempting to explain the discrepancy between calculated and experimental values either in terms of specific adsorption (higher affinity) of one or the other of the lattice ions or in terms of lack of equilibrium conditions in the experiments, but we are going to refrain from making such explanations at the present. An important feature of the solid/solution equilibrium is that the predominant charged species at pH's less than about 8 is Ca^{++} whereas it is OH^- at pH's greater than 9. Thus, the positive charge is probably due to preferential adsorption of Ca^{++} ions whereas the negative charge results from the adsorption of OH^- ions.

In several natural or laboratory conditions involving apatite, only hydrogen ion concentration is changed without the addition of any soluble salts of calcium or phosphorous. The concentration of calcium and phosphorous species is determined by solid/solution equilibrium. The line of zero charge in such cases becomes the point of zero charge and will be designated as PZC(pH). The values of PZC(pH) determined by various investigators are listed in Table 4. It can be seen from this table that significant variations in the surface charge characteristics of apatites have been observed. Although substituted impurities and lattice defects are often cited as the reasons for these variations, a clear understanding of the role of impurities on the surface charge of apatites is largely lacking.

Calcium ions exhibit a complex behavior towards their effect on the surface charge of apatites. These ions act as potential-determining ions as well as specifically adsorbed ions, as demonstrated by Mishra et al. (45) and Somasundaran (48).

UPTAKE OF FLUORIDE IONS

The chemical aspects of the uptake of fluoride by hydroxyapatite have been studied by many researchers and reviewed recently (49). The following are some of the proposed mechanisms:

(a) OH^-/F^- Exchange on Hydroxyapatite. McCann (50) and Leach (51) investigated the reactions of fluoride with synthetic hydroxyapatite and powdered enamel, respectively. They both arrived at the conclusion that at low fluoride concentrations fluorapatite was the sole product of the reaction. They postulated that fluorapatite was formed by substitution of F^- ions for OH^- ions in the apatite lattice structure as shown in Equation 1.

$$Ca_{10}(PO_4)_6(OH)_2 + 2\ F^- = Ca_{10}(PO_4)_6F_2 + 2\ OH^- \qquad (1)$$

The reaction was considered to occur mainly at the crystal surfaces. Saleeb and deBruyn(19), on the basis of electrophoretic

Table 4. Point of Zero Charge or Isoelectric Point of Apatites in
the Absence of Added Calcium or Phosphate Salts

Apatite	PZC or IEP, pH	Method	Reference
Hydroxyapatite (synthetic)	8.5	Titration	44
	8.5	Electrophoresis	19
	6.5	Electrophoresis	45
	6.8	Electrophoresis	52
Fluorapatite (natural)	5.6	Streaming Poten.	48
	4	Streaming Poten.	52
	6	Streaming Poten.	52
	6.9	Titration	44

measurements, concluded that F^- readily exchanges with OH^-. This
ion exchange was found to be irreversible, however(3). Most of the
change in electrophoretic mobility occurred in the fluoride ion
concentration range of 10^{-6} to 10^{-5} M and no change was observed
in the concentration range of 10^{-5} to 10^{-3} M. Increase in the
absolute value of the zeta potential of hydroxyapatite was observed
by several authors (51,53). The increase was explained by Chander
and Fuerstenau (53) in terms of an increase in the surface charge
density of the solid when fluoride ions are adsorbed. The substitution of hydroxyl ions by fluoride ions is ascribed to a better
fit of the fluoride ions in the hydroxyapatite lattice. Several
investigators have failed to observe a direct correspondence
between the fluoride ions adsorbed and hydroxyl ions released
(19,53). Chander and Fuerstenau (52) observed that the uptake of
fluoride ions is accompanied by an uptake of calcium ions. Thus,
it is likely that the fluoride ion uptake occurs both by specific
adsorption and by ion exchange mechanisms. At higher
concentrations of fluoride ions (> 100 ppm) the exchange was
accompanied by the partial disintegration of the crystal lattice
and the formation of CaF_2.

$$Ca_{10}(PO_4)_6(OH)_2 + 20\ F^- = 10\ CaF_2 + 6PO_4^{3-} + 2OH^- \qquad (2)$$

The fluoride ion concentration at which CaF_2 forms is shown by an

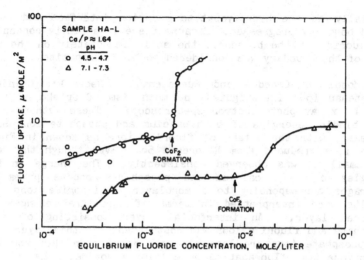

Fig. 3. The uptake of fluoride ions on hydroxyapatite as a function of fluoride ion concentration and pH (49).

arrow in Figure 3 which is reproduced from Reference 49 with a slight modification. At concentrations just below the CaF_2 formation, the fluoride uptake remains constant independent of fluoride concentration. More recently, similar results are reported by Yesinowski, et al. (54). In this concentration range where the adsorption isotherm is parallel to the concentration axis, structural changes in the adsorbed layer are postulated (see Mechanism (e) below.)

(b) Model of Spinelli et al. The mechanism proposed by Spinelli et al. (54) for fluoride uptake by hydroxyapatite involves three processes: precipitation of fluorapatite; dissolution of the solid hydroxyapatite followed by precipitation of fluorapatite (recrystallization); and removal of fluoride ions by adsorption and ion exchange.

(c) Model of Ramsey et al. Ramsey et al. (56) studied the uptake of fluoride by hydroxyapatite as a function of solution pH and concluded that the effect of changing the pH of the fluoride solution is to increase the F^- content in the hydroxyapatite by partial dissolution of hydroxyapatite at pH 4.0, leading to the formation of CaF_2 and transformation of CaF_2 into fluorapatite as the pH increases.

(d) Theoretical model of Nelson and Higuchi. Nelson and Higuchi derived a mathematical model based on diffusion and chemical reactions at moving boundaries (57). They suggested that CaF_2 forms initially at the hydroxyapatite solution interface and that the solid boundary progresses. Because there is a finite concentration of fluoride at the boundary, the solid immediately on the apatite side of the boundary was considered to be fluorapatite.

(e) Model of Chander and Fuerstenau. Recently, Chander and Fuerstenau (58) investigated the mechanism of uptake of fluoride ions by x-ray photoelectron spectroscopy. These authors observed that binding energies of both calcium and phosphate ions slightly decrease, with the uptake of fluoride ions as shown in Figure 4, which is reproduced from Reference 58. Even though the decrease was small it was observed consistently. The change in binding energies of both calcium and phosphorus occurs primarily at coverages corresponding to a monolayer of fluoride ions. These results were interpretted in terms of structural changes in the adsorbed layer. An interfacial layer consisting of calcium, phosphate and fluoride ions was postulated. In this layer, calcium and phosphate ions are less strongly bound than they would be in hydroxyapatite, fluorapatite or calcium fluoride. The interfacial layer possibly grows into multilayers and might be the precursor to nucleation of a new crystalline phase, fluorapatite at low fluoride concentrations or calcium fluoride at high fluoride concentrations. Yesinowski et al. (54) also observed a surface layer which has NMR spectra in between that of CaF_2 and fluorapatite. The spectra was observed to change on aging. Yesinowski et al. interpreted their results in terms of the formation of a surface layer of fluorohydroxyapatite which transforms to fluorapatite on aging. These investigators neither provide information regarding the driving force for converting fluorohydroxyapatite to fluorapatite nor the mechanism of the transformation. We consider that the interfacial layer forms through dissolution of hydroxyapatite and reprecipitation in fluoride solutions. On aging, this layer transforms into more stable fluorapatite or calcium fluoride.

Chander and Fuerstenau (53) studied the mechanism of uptake of stannous fluoride by hydroxyapatite with electrokinetic and solubility measurements. These authors observed that under comparable fluoride concentrations, the mobility of HA is influenced much more strongly by SnF_2 than by NaF. Stannous ions were postulated to chemisorb through formation of a $SnF_x(PO_4)^{2-x-3y}$ surface complex according to the reaction:

$$y \ (PO_4)_{surf} + (SnF_x^{2-x})_{aq} = (SnF_x(PO_4)_y^{2-x-3y})_{surf} \qquad (3)$$

The solubility of HA could be interpreted in terms of the change in F^- concentration alone. Although stannous ions were strongly chemisorbed, they have an insignificant effect on the solubility.

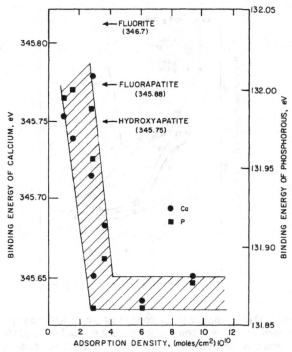

Fig. 4. The binding energies of calcium and phosphorous in fluoride
treated hydroxyapatite, plotted as a function of fluoride
adsorption density. The binding energies of calcium in
calcium fluoride, fluorapatite and hydroxyapatite are shown
by the arrows (58).

ADSORPTION OF ORGANIC IONS

Interest in understanding the nature of interactions between
short chain organic surfactants and large molecular weight macromolecules and ions with hydroxyapatite extends in several fields.
In the area of caries prevention and control, surfactant adsorption
plays an important role in the initial stages of plaque formation
and in the adhesion of tooth restorative materials to the surface
of a freshly prepared cavity (59,60). Interaction of polypeptides
in human urine with hydroxypaptite is important in human biology as
hydroxyapatite has been found as a major or minor component in a
majority of kidney stones (61). Flotation separation of apatite
from gangue minerals is an important industrial operation in which
surfactants are used to effect separations (62). Dewatering of
colloidal phosphatic slimes which are generated in large quantities
in the processing of phosphate rock is a major industrial problem
which has been studied by a number of researchers in recent years

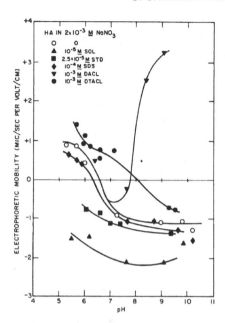

Fig. 5. The electrophoretic mobility of hydroxyapatite in the presence of various surfactants: SOL - sodium oleate; STD - sodium tridecanoate; SDS - sodium dodecyl sulfonate; DACl - dodecylammonium chloride; DTACl - dodecyltrimethylammonium chloride. Ionic strength: 2×10^{-3} M $NaNO_3$.

(63). Only a few systematic studies have been carried out to delineate the mechanism of interaction of organic surfactants and macromolecules with hydroxyapatite. Mishra, et al. (45) studied the effect of carboxylic acids (oleic and tridecanoic acids), dodecylsulfonate and amines (dodecylammonium chloride and dodecyltrimethylammonium chloride) on the electrophoretic mobility of hydroxyapatite. Their representative data are plotted in Figure 5. These authors considered that carboxylate ions interact with calcium ions to form calcium oleate or calcium tridecanoate at the surface. Since the pK_{sp} of the two salts are quite comparable (pK_{sp} = 12.4 for calcium oleate and 13.5 for calcium tridecanoate), the higher surface activity of oleate ions was attributed to their longer hydrocarbon chains. The hydrocarbon chains of adsorbed molecules can associate to form hemimicelles at the interface, thereby increasing the free energy of adsorption.

Sulfonate ions also exhibit surface activity as seen by the shift in the point of zeta potential reversal (PZR). These ions show much less surface activity than the carboxylate ions which is expected because of the relatively smaller affinity of sulfonate ions for calcium ions (the pK_{sp} of calcium dodecyl sulfonate being 11.7). At pH's greater than 9 or so, tridecanoate ions are ineffective because of the precipitation of calcium tridecanoate in the bulk solution. Recently, Voegel et al. studied the release of phosphate and calcium ions during adsorption of benzene polycarboxylic acids onto apatites (64). Juriaanse et al. observed a similar release of calcium and phosphate ions during the adsorption of polypeptides on dental enamel (65). These results demonstrate that complex interfacial and dissolution interactions occur in these systems.

Specific adsorption of dodecylammonium and dodecyltrimethyl-ammonium ions on hydroxyapatite was observed by Mishra et al. (45). The dodecylammonium ions formed hemimicelles through the association of hydrocarbon chains, with possible co-adsorption of neutral amine molecules. The large head group of dodecyltrimethyl-ammonium ions prevents the chains from coming together, thus inhibiting hemimicelle formation. Similar to the adsorption of amines on hydroxyapatite is the adsorption of amines groups in basic polypeptides on dental enamel. Specific adsorption of amines may occur through hydrogen bonding between the amine ions and the phosphate groups of hydroxyapatite (66) although no confirmatory data are available.

Adsorption of polyphosphonate at the water/hydroxyapatite interface has been studied by Rawls et al. (67). It was found that phosphate ions are released into solution in molar amounts exceeding the amount of phosphonate groups adsorbed. The liberation of phosphate ions continued long after the polymer adsorption density had become constant, indicating changes in the interfacial region occur even if the polymer adsorption has reached an apparent equilibrium. Although the authors explained their observations in terms of the rearrangement of adsorbed polymer the interaction of calcium ions with the polymer cannot be ruled out. An excess of calcium ions is expected in the interfacial region if phosphate ions are released into the bulk solution.

Only a few studies have been carried out to quantitatively determine the parameters of adsorption of surfactant and macro-molecular ions on apatites. A summary of the results is given in Table 5 for both hydroxyapatite and fluorapatite. In most cases, Langmuir-type adsorption isotherms fit the observed data and therefore the Langmuir parameters are given.[*] Figure 6 gives the results of glutamic acid and lysine adsorption on hydroxyapatite. The adsorption of glutamic acid shows a chemisorption mechanism whereas lysine adsorbs through electrostatic interactions (the PZC pH of this hydroxyapatite preparation was pH = 6.5). The isotherms

---

[*]see Appendix I

Table 5. Parameters for the Adsorption of Surfactants and
Macromolecules on Apatites.

Langmuir Adsorption Parameters

Apatite	Adsorbate (Molecular Weight)	K	Γ_m μ mole/m^2	Ref.
Hydroxy-apatite	L-Aspartic acid (133)	7.27x10^2*	21.6***	68
		3.0x10^2*	0.44	69
	L-Glutamic acid (147)	3.19x10^3*	18.94***	68
		5.2x10^4	0.128	70
	Separan 10 (71)**	1.39x10^4	3.38	71
	Bovine albumin (6.9x10^4)	2.49x10^7*	0.028	69
	Benzenehexa-carboxylic acid	18.6*	41	72
	Hexa sodium benzenehexa-carboxylate	32.8*	8.5	72
	Benzene 1,3,5,-tricarboxylic acid	0.74*	74	72
	L-Arginine	Non-Langmuirian Ads.		73
Fluor-apatite	L-Aspartic acid (133)	3.76x10^3*	0.68	69
	Separan 10 (71)**	1.39x10^4	3.38	71
	Bovine albumin (6.9x10^4)	4.49x10^7*	0.041	69

* recalculated from data
** monomer
*** mole/g

for the adsorption of both glutamic acid and lysine do not follow
Langmuir equation and those results will be presented
elsewhere(74). These and the earlier studies reviewed by Rolla
(75) show that the uptake of surfactant ions and macromolecules can
occur in several ways. The following are some of the proposed
mechanisms:

(a) Physical adsorption in the Electrical Double Layer. Adsorption
    occurs through electrostatic interactions between the apatite
    surface and the adsorbing species. Interactions between the

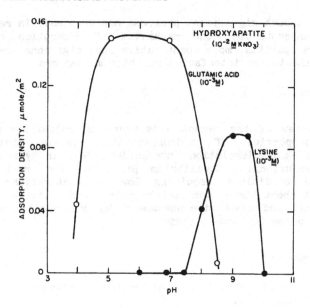

Fig. 6. The adsorption of glutamic acid and lysine on hydroxy-
        apatite at various pH's.

adsorbed molecules may lead to hemimicelle formation as
observed, for example, by Mishra et al. (45) for the adsorption
of dodecylamine.

(b) <u>Chemisorption</u>. Chemisorption of anions on cationic sites and
cations on anionic sites may occur thorough direct bond formation between the adsorbing species and the apatite surface.

(c) <u>Binding through Specifically Adsorbed Calcium Ions</u>.
Specifically adsorbed calcium ions may act as a bridge to bind
anionic species to phosphate sites of the apatite lattice.

(d) <u>Exchange of Lattice Ions</u>. The adsorbing species may exchange
with the lattice ions of the solid.

(e) <u>Binding through Hydrogen Bonds</u>. The adsorbing species may
hydrogen bond to the phosphate or hydroxyl (or fluoride) groups
of the apatite lattice.

(f) <u>Formation of Multilayers through Surface Reaction</u>. Multilayers
are formed particularly when plaque forming proteins interact
with hydroxyapatite surface.

Even though significant progress has been made in recent years towards understanding the mechanism of adsorption of organic species on apatites, more quantitative investigations are needed to fully delineate the interfacial reaction mechanisms.

SUMMARY

A review of the recent literature on solubility and inter-facial properties of hydroxyapatite reveals that complex interfacial and dissolution interaction occur in hydroxyapatite/aqueous solutions. Equilibrium prevails only under carefully controlled conditions requiring low supersaturations and long times. At short times, non-stoichiometric and metastable phases at the surface, and adsorption phenomena may dominate the behavior of the solid phase in these systems.

ACKNOLWEDGMENTS

The authors acknowledge the National Science Foundation for support of this work.

REFERENCES

1. V. K. La Mer, J. Phys. Chem. 66:973 (1962).
2. W. E. Brown, "Environmental Phosphorous Handbook," G. Beeton, ed., Wiley, New York, Ch. 10 (1973).
3. C. L. Kibby and W. K. Hall, "Chemistry of Biosurfaces," Vol. 2, M. L. Hair, ed., Marcel Dekker, New York, Ch. 15 (1972).
4. L. E. Holt, V. K. La Mer and H. B. Chown, J. of Biol. Chem. 64:509 (1925).
5. H. M. Rootare, V. R. Dietz and F. C. Carpenter, J. Colloid Sci. 17:179 (1925).
6. N. Bjerrum, "Investigations on the Solubility of Calcium Phosphates - Selected Papers," Einar Munksgaard, Copenhagen, 1949, cited in Reference 21.
7. K. L. Elmore, R. Kunin, C. M. Mason and J. D. Hatfield, T.V.A. Chemical Engineering Report no. 8 (1950), cited in Ref. 21.
8. J. S. Clark, Canadian J. Chem., 33:1969 (1955).
9. G. L. Levinskas and W. F. Neuman, J. Phys. Chem. 59:164 (1955).
10. E. Hayek, W. Boeher, J. Lechleitner and H. Petter, Z. Anorg. Allg. Chem. 295:241 (1958).
11. W. E. Brown, Soil Sci. 90:51 (1960).
12. W. L. Lindsay and E. C. Moreno, Proc. Soil Sci. Soc. Amer. 24:177 (1960).
13. S. R. Olsen, F. S. Watanabe and S. V. Cole, Trans. 7th Int. Congr. Soil Sci. 2:397 (1960).
14. A. R. Hagen, "On the Behavior of Dental Enamel in Organic Salt Solutions," Scandinavian University Books, Oslo (1965).

15. H. W. Fassbender, H. C. Lin and B. Ulrich, Z. Pfl. Ernaehr. Dueng. Bodenk. 112:101 (1966).
16. E. C. Moreno, T. M. Gregory and W. E. Brown, J. Res. Nat. Bur. Stand. 72A:773 (1968).
17. D. R. Wier, S. H. Chien and C. A. Black, Soil Sci. 111:107 (1971).
18. S. H. Chien, "Ion Activity Products of Some Apatite Minerals," Ph.D thesis, Iowa State University, (1972).
19. F. Z. Saleeb and P. L. de Bruyn, J. Electroanal. Chem. 37:99 (1972).
20. Y. Avnimelech, E. C. Moreno and W. E. Brown, J. Res. Nat. Bur. Stand. 77A:149 (1973).
21. A. N. Smith, A. M. Posner and J. P. Quirk, J. Colloid Interface Science 54:176(1976).
22. M. S. Wu, W. I. Higuchi, J. L. Fox and M. Friedman, J. Dent. Res. 55:496 (1976).
23. W. E. Brown, T. M. Gregory and L. C. Chow, Caries Res. 11(Suppl. 1) :118 (1977).
24. H. McDowell, T. M. Gregory, and W. E. Brown, J. Res. Nat. Bur. Stand. 81A:273 (1977).
25. M. B. Fawzi, Z. L. Fox, M. G. Dedhiya, W. I. Higuchi and J. J. Hefferren, J. Colloid Interface Science. 67:304 (1978).
26. S. Larsen, Nature 212:605 (1966)
27. T. P. Feenstra and P. L. de Bruyn, J. Phys. Chem. 83:475(1979).
28. A. G. Walton, W. J. Bodin, J. Fueredi and A. Schwartz, Canadian Journal of Chemistry 45:2695 (1967).
29. A. S. Posner, Physiological Review 49:760(1969).
30. H. Neweseley, Arch. Oral. Biol., Spec. Suppl. 6:174 (1961).
31. W. E. Brown, J. P. Smith, J. R. Lehr and A. W. Frazier, Nature, 196:1048 (1962).
32. G. M. Noncollas and B. Tomazic, J. Phys. Chem. 78:2218(1974).
33. M. D. Francis and N. C. Webb, Calcif. Tissue Res. 6:335 (1971).
34. A. L. Boskey and A. S. Posner, J. Phys. Chem.. 80:40(1976).
35. W. G. M. van den Hock, T. P. Feenstra and P. L. de Bruyn, J. Phys. Chemistry. 84:3312(1980).
36. Z. Amjad, P. G. Koutsoukos and G. H. Nancollas, J. Colloid Interface Sci., 82:394 (1981).
37. S. Chander, C. C. Chiao and D. W. Fuerstenau, J. Dental Res. 61(2):403 (1982).
38. S. Chander and D. W. Fuerstenau, J. Colloid Interface Sci. 70:506 (1979).
39. G. A. Parks, Chem. Reviews 65:177 (1965).
40. G. A. Parks, A.C.S. Adv. Chem. Series 67:121 (1967).
41. E. P. Honig and J. H. Th. Hengst, J. Colloid Interface Sci. 29:510(1969).
42. R. J. Roman, M. C. Fuerstenau and D. C. Seidel, Trans. AIME 241 :56(1968).
43. J. D. Miller and J. B. Hiskey, J. Colloid Interface Sci. 41:567 (1972).

44. L. C. Bell, A. M. Posner and J. P. Quirk, J. Colloid Interface Sci. 42:250(1973).
45. R. K. Mishra, S. Chander and D.W. Fuerstenau, Colloids and Surfaces 1:105(1980).
46. S. K. Mishra, Int. J. Mineral Processing, 5:69 (1978).
47. C. C. Chiao, "Transformation of fluorite to fluorapatite in phosphate solutions," Ph.D thesis, University of California, Berkeley, 1976.
48. P. Somasundaran, J. Colloid Interface Sci. 27:659(1968).
49. J. Lin, S. Raghavan, and D. W. Fuerstenau, Colloids and Surfaces, 3:357 (1981).
50. H. G. McCann, J. Biol. Chemistry 201:247(1953).
51. S. A. Leach, Brit. Dental J. 106:133(1959).
52. P. Somasundaran and G. E. Agar, J. Colloid Interface Sci. 24:433(1967).
53. S. Chander and D. W. Fuerstenau, Colloids and Surfaces 4:101(1982).
54. J. P. Yesinowski, R. A. Wolfgang, M. J. Mobley, this publication.
55. M. A. Spinelli, F. Brudevold and E. C. Moreno, Arch. Oral. Biol. 16:187(1971).
56. A. C. Ramsey, E. J. Duff, L. Paterson and J. L. Stuart, Caries Res. 7:231(1973).
57. N. G. Nelson and W. I. Higuchi, J. Dent. Res. 49:1541(1970).
58. S. Chander and D. W. Fuerstenau, An XPS study of the Fluoride Uptake by Hydroxyapatite, submitted for publication in Colloids and Surfaces,1983.
59. R. P. Quintana, "Applied Chemistry at Protein Interface," Ed. R. E. Baier, Adv. Chem. Series 145:290(1975).
60. D. I. Hay, Arch. Oral Biol. 12:937 (1967).
61. R. S. Malek and W. H. Boyce, J. Urol. 117:336(1977).
62. P. R. Smith, Jr., "Flotation, A. M. Gaudin Memorial Volume," Ed. M. C. Fuerstenau, AIME, New York, Vol. 2, 1265 (1976).
63. D. R. Nagraj, L. McAllister and P. Somasundaran, Int. J. Min. Proc., 4:111(1977).
64. J. C. Voegel, S. Gillmeth and R. M. Frank, J. Colloid Interface Sci. 84:108 (1981).
65. A. C. Juriaanse, J. Arends and J. J. Ten Bosch, J. Colloid Interface Sci. 76:220 (1980).
66. A. C. Juriaanse, J. Arends, and J. J. Ten Bosch, J. Colloid Interface Sci., 76:212 (1980).
67. H. R. Rawls, T. Bartels and J. Arends, J. Colloid Interface Sci. 87:339 (1982).
68. J. V. Garcia-Ramos, P. Carmona and A. Hidalgo, J. Colloid Interface Sci. 83:479 (1981).
69. E. C. Moreno, M. Kresak and R T. Zahradnik, Caries Res. 11(Suppl.1):142 (1977).
70. J. Lin and D. W. Fuerstenau, unpublished results.
71. Pradip, Y. A. Attia and D. W. Fuerstenau, Colloid & Polymer Science, 258:1343 (1980).

72. J. C. Vogel and R. M. Frank, J. Colloid Interface Sci. 83:26(1981).
73. J. V. Garcia-Ramos and P. Carmona, Canadian J. Chem. 59:222 (1981).
74. D. W. Fuerstenau, J. Lin, S. Chander and G. D. Parfitt, "Adsorption and Electrokinetic Effects of Amino Acids on Rutile and Hydroxyapatite," paper to be presented at the ACS Meeting, Washington, D. C., Aug. 28 - Sept. 2, 1983.
75. G. Rolla, Caries Research 11(Suppl.1):243 (1977).

APPENDIX I. LANGMUIR ISOTHERM FOR ADSORPTION FROM SOLUTION

We have observed that there is some confusion about the use of the Langmuir adsorption isotherm in the dental and biological literature, particularly with reference to the units of the parameter K. A brief derivation is given below for the benefit of researchers who may not be familiar with the Langmuir isotherm. For adsorption from solution, equilibrium can be described as follows:

Solute (soln, a_2) + Solvent (ads, a_1^s)

= Solute (ads, a_2^s) + Solvent (soln, a_1)

where a_1 and a_2 are activities of solvent (water) and solute (surfactant or polymer) in solution and a_1^s and a_2^s are the respective activities at the surface. The equilibrium constant, K, is given by:

$$K = \exp\left(-\Delta G^o_{ads}/RT\right) = (a_2^s)(a_1) / (a_1^s)(a_2)$$

where ΔG^o_{ads} is the free energy change for adsorption. K is therefore a dimensionless quantity and is related to the free energy of adsorption by the above equation. For dilute solutions, the activity of solvent is unity and that of solvent can be replaced by mole fraction. The activity on the surface can be assumed to be proportional to the adsorption density. Then,

$$K = \exp\left(-\Delta G^o_{ads}/RT\right) = \Gamma / X_2(\Gamma_m - \Gamma)$$

where X_2 is the mole fraction of solute, Γ is the adsorption density, and Γ_m is the adsorption density at saturation. If the concentration of solute, C, is in moles per liter

$$K = \Gamma / [55.5 \, C \, (\Gamma_m - \Gamma)]$$

where the term 55.5 equals the number of moles of water per liter solution.

INFLUENCE OF DODECYLAMINE HYDROCHLORIDE ADSORPTION ON THE DISSOLUTION KINETICS OF HYDROXYAPATITE

P.W. Cho, J.L. Fox, W.I. Higuchi* and P. Pithayanukul

Department of Pharmaceutics, College of Pharmacy
University of Utah, Salt Lake City, Utah, U.S.A.
*To whom correspondence should be addressed

ABSTRACT

The adsorption of dodecylamine (DAC) onto hydroxyapatite (HAP) crystal surfaces and its influence on the dissolution rate behavior of HAP crystal suspensions have been studied. The sigmoidal nature of adsorption isotherms indicates that there was considerable cooperativity among the DAC molecules in the adsorbed state leading to patchwise adsorption of "islands" which could completely stop the dissolution from the HAP surface beneath them. This mechanism is consistent with the observation that there was an almost quantitative, simple inverse type relationship between the adsorption and the dissolution rate of HAP in an acidic media. At the highest DAC concentration and maximum DAC adsorption, there is a residual dissolution rate component which is about 15% of the zero DAC rate. It is suggested that this residual rate may be due to either incomplete adsorption because maximum adsorption occurs near the critical micelle concentration or due to the presence of another dissolution site(s) on the HAP crystal that is insensitive to rate inhibition by DAC.

INTRODUCTION

Recently, we have been able to correlate the dissolution behavior of compressed hydroxyapatite (HAP) pellet in acidic media with crystal suspensions. The dissolution has been interpreted based on dissolution from two or more sites[1,2]. In the studies with HAP suspensions, the analysis of the problem also involved showing under what condition dissolution is essentially 100% surface controlled.

51

It is the purpose of this study to investigate the influence of a long chain alkyl amine, dodecylamine (DAC), on the dissolution rate of HAP. Previously, Roseman et al.[3] had shown that DAC at millimolar levels may significantly inhibit the dissolution of both synthetic HAP and human dental enamel in acidic media. However, in Roseman's studies no attempt was made to correlate the adsorption of DAC on the HAP surface and the dissolution in a quantitative way and contributions from the diffusion and the surface controlled kinetics were not delineated. It has been our hope that using the well-defined HAP suspension method, we should be able to discern a relationship between DAC adsorption and surface controlled dissolution kinetics.

MATERIALS

Hydroxyapatite

Samples used in this study were labeled as UM-T, UM-K, UM-A, UM-M, NBS-P, Y146, Y155, Bio-Gel and Calbiochem.

The procedures for preparing NBS-P and UM-T were described in detail in a previous paper[2]. UM-K was prepared like UM-T except that the digestion time was shortened to one month

Both Y146 and Y155 were provided by Dr. R.A. Young of Georgia Institute of Technology. Y155 was hydrothermally prepared in 2 steps as shown below:

1. $4CaO + 3Ca_2P_2O_7 \xrightarrow[\text{dry } N_2]{1000\ ^0C} 10CaO \cdot 3P_2O_5$

2. $10CaO \cdot 3P_2O_5 \xrightarrow[\text{15000 psi } H_2O,\ 1\ \text{month}]{500\ ^0C} Ca_{10}(PO_4)_6(OH)_2$

Y146 was prepared by boiling a dicalcium phosphate dihydrate solution for 1 month. It was a defect hydroxyapatite with formula:

$$Ca_{(10-x)}(PO_4)_{(6-x)}(OH)_{(2-x)}(HPO_4)_{(x)}$$

UM-A and UM-M, which were the major samples used in this investigation, were synthesized in this laboratory by the same method. It was a slight modification of the one described by Avnimelech[4]. Pure phosphoric acid was added at a slow controlled rate to a boiling, carbon dioxide free, calcium oxide solution in a Teflon[R] flask. The resulting precipitate was digested for 2 weeks in the boiling reacting solution with N_2 constantly

bubbling. The residue was centrifuged and the supernatant liquid
was decanted. After fresh boiling double distilled water was
added, it was stirred and centrifuged again. The same procedure
was repeated three times. The final residue was then dried at
110°C overnight.

The two commercially available samples, Bio-Gel and
Calbiochem, were obtained from Bio-Gel Laboratories and
Calbiochem-Behring Corp., respectively.

The chemical compositions and specific surface areas of the
samples are given in Table 1.

Table 1. The Chemical Composition, Specific Surface Area
and Major Impurities of the HAP Samples

HAP	Ca/P[a] (Molar ratio)	Specific Surface Area[b] (m^2/gm)	Major Impurities
NBS-P	1.714	22.5	none
UM-T	1.664	-	0.3% CO_3
UM-K	1.732	22.9	1.3% SiO_2
UM-A	1.703	23.4	0.2% CO_3
UM-M	1.692[c]	23.0	none
Y146	1.60	-	HPO_4
Y155	1.65	2.3	none
Bio-Gel	-	69.8	not analyzed

[a]Ca/P molar ratio determined by General Electric Co. (Cleveland,
Ohio); calcium by an EDTA volumetric method and phosphate by a
triple precipitation gravimetric method.
[b]Specific surface area determined by Micromeritics (Norcross,
Georgia) based on B.E.T. calculations using nitrogen adsorption
data.
[c]Determined by Galbraith Lab. (Knoxville, Tennessee) and General
Electric Co. (Cleveland, Ohio).

Dodecylamine Hydrochloride (DAC)

Then Reagent grade DAC was obtained from Eastman Kodak
Company. It was recrystallized once in benzene. The purity was
verified by means of paper chromatography.

Solutions

        Calculated amounts of acetic acid and sodium acetate stock
solutions were used to make up the buffer solution of desired pH
which, in this study, was 4.50. DAC solutions were made by
dissolving predetermined amounts of DAC in the buffer. Sodium
chloride was added to all buffer solutions to maintain the ionic
strength of 0.1. The pH of the solutions were measured to an
accuracy of ± 0.01 pH unit using an Altex model 4500 digital pH
meter. Reagent grade chemicals were used in all preparations.

EXPERIMENTAL METHODS

Dissolution

        Hydroxyapatite in clumps was broken down by light grinding
until no visible lumps were present. Exactly 10 mg of HAP was
suspended in 12.5 ml of double distilled water in a 50 ml water-
jacketed (30°C) reaction vessel. This suspension was magnetically
stirred for around 20 sec at 600 rpm before being sonified
(Ultrasonic Generator Model AK500, Acoustica, Los Angeles, CA) for
60 sec to produce a milky suspension. After sonication, stirring
was resumed and 12.5 ml of dissolution medium (double
concentration) was added to the suspension. Addition and mixing
was completed in 1 to 2 sec, and the pH at this point was recorded
to be 4.50 ± 0.01. Samples of 0.5 ml were withdrawn at
predetermined times by an Eppendorf pipet during the dissolution
experiment and immediately passed through a Millipore filter paper
(GSU) with a pore size of 0.22μ. Sampling and filtering was
completed within 4 to 6 sec. The filtrate was analyzed for
phosphate and/or calcium.

Adsorption

        Five hundred mg of HAP was transferred to a 25 ml water-
jacketed (30°C) vessel containing 5 ml of double distilled
water. The suspension was sonified for about 3 min before
starting the magnetic stirrer (600 rpm). Five ml of DAC solution

(double concentration) was then added to the suspension. Addition and mixing was completed in around 2 sec. The system was kept closed throughout the experiment to avoid evaporation. Samples of 0.5 ml were withdrawn at predetermined times using an Eppendorf pipet and then filtered through Millipore filter paper (GSU) with a pore size of 0.22μ. Sampling and filtering was completed within 4 to 6 sec. The filtrate was analyzed for phosphate and dodecylamine hydrochloride.

ANALYTICAL METHODS

Phosphate

Phosphate concentrations were determined according to the method of Gee et al.[5] The phospho-ammonium molybdate complex formed was reduced by stannous chloride. The absorbance of the resulting color was determined after 15 minutes at $\lambda = 720$ nm in a Beckman Model 25 spectrophotometer.

Dodecylamine Hydrochloride

DAC concentrations were determined according to the method of Ino et al.[6] with slight modifications. This antagonistic titration used sodium lauryl sulfate as the titrant and rhodamine 6G as the adsorption indicator. The simultaneous formation of a pink precipitate and color change, with loss of fluorescence, was taken to be the equivalence point.

RESULTS

The Determination of the Critical Micelle Concentration (CMC)

The CMC of DAC was determined by measuring the surface tension as a function of the surfactant concentration using a Rosano Surface Tensiometer. The point beyond which the surface tension became constant was taken to be the CMC (Fig. 1). As indicated by an arrow in Fig. 1, the CMC for DAC dissolving in the medium (0.1 M acetate buffer, pH 4.50, $\mu = 0.1$) which was used throughout this study was about 3.5 mM.

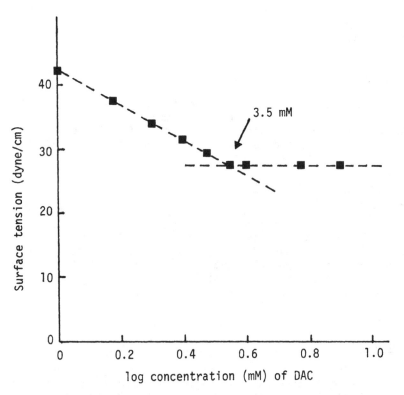

Fig. 1   Surface tension vs. the logarithm of DAC concentration (mM)
          in 0.1 M acetate buffer, pH 4.50, μ=0.1 at 30° C.

Dissolution

Preliminary studies showed that the dissolution did not depend on the preadsorption time. Experiments conducted by equilibrating the HAP powder with DAC for 0, 15, 30 and 60 min before the addition of the acidic buffer all provided essentially identical results. This observation suggested that the adsorption process was relatively fast. Adsorption rate studies which will be discussed in the following section also seemed to support this view.

As in the baseline work done in the absence of DAC, all plots of the amount of HAP dissolved versus time exhibited a biphasic pattern characterized by a rapid initial step for approximately 15 minutes followed by a region of very little dissolution. Figure 2 shows all the raw data for the sample UM-M with variable amounts of DAC in the medium. For DAC concentrations up to 4 mM, increasing DAC resulted in a decrease of dissolution although the decrease was far from linear. No significant change could be seen beyond this concentration. The curves of 4, 4.5, 5 and 6 mM DAC were not distinguishable. Further increase of DAC in the medium was impossible due to the low solubility of DAC in the buffer solution. The initial dissolution rate, J, was estimated from the linear region. Figure 3 is a plot of J as a function of DAC in the medium. It is interesting to note that the drop of dissolution rate with increasing DAC is not linear. Rather, there is a large sudden drop of rate between 1.5 and 2.5 mM DAC. In this region, a small change of DAC concentration reduces the dissolution rate to about 15% of the baseline value. The same trend was also observed in samples UM-T, UM-K, UM-A, NBS-P and Y155. The data are presented in Fig. 4. Qualitatively they all show this unusual behavior although the magnitude of drop is different from sample to sample. Despite the differences in the rates, about a 50% drop in the rate occurs at about 2.0 mM DAC for all samples (Table 2).

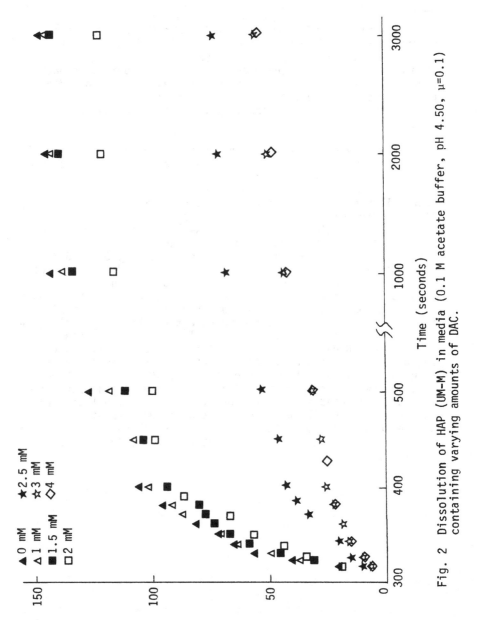

Fig. 2 Dissolution of HAP (UM-M) in media (0.1 M acetate buffer, pH 4.50, μ=0.1) containing varying amounts of DAC.

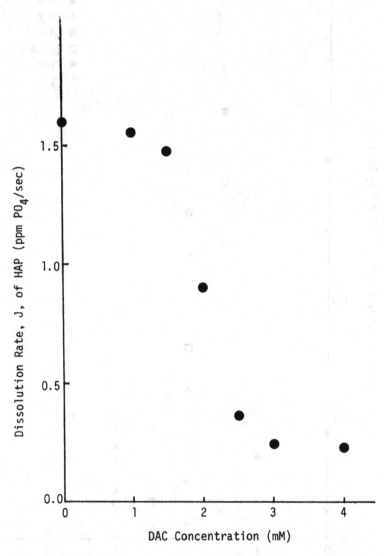

Fig. 3   Dissolution rate, J, of HAP (UM-M) as a function
of DAC in the medium (0.1 M acetate buffer, pH
4.50, μ=0.1).

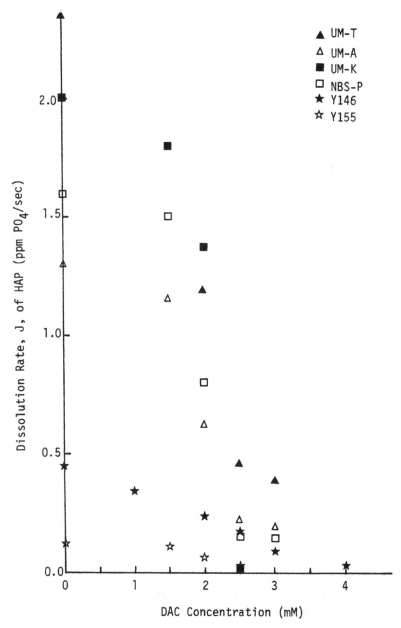

Fig. 4   Dissolution rate, J, of different HAP as function
of DAC in the medium (0.1 M acetate buffer, pH 4.50,
μ=0.1).

Table 2.   Concentration of DAC at which the Dissolution Rate is 50% of Control

HAP Sample	DAC Concentration (mM) at 50% $J_{control}$
NBS-P	2.11
UM-T	2.10
UM-A	2.08
UM-M	2.15
UM-K	2.25
Y146	2.16
Y155	2.09

Sample Y146, a preparation known to contain HPO_4 as an impurity, exhibited a pattern that differs somewhat from that of the others. As presented in Fig. 4, although a 50% drop of rate also occurs at about 2.0 mM, the dependence of dissolution rate inhibition upon DAC concentration is much more gradual than for the other preparations. The data indicate that Y146 is an outlier.

Adsorption

Adsorption rate studies revealed that adsorption occurred very rapidly. Analysis of the sample taken 5 seconds after the addition of the DAC solution indicated that adsorption equilibrium was reached in approximately 10 seconds. During this process no more than 1% of the starting material dissolved in the medium and pH rose from 4.5 to approximately 4.8.

Due to the limited supply of most of the samples, only UM-A and UM-M were studied extensively. The results of adsorption experiments are presented in Fig. 5. It can be seen that the curves are more or less sigmoidal and deviate significantly from Langmurian behavior. The same kind of sigmoidal isotherms were also observed in the two commercial samples, Bio-Gel and Calbiochem (Fig. 6). The levels of adsorption for UM-M and the commercial samples differed by about a factor of three which is consistent with the specific surface area difference between the UM-M and the commercial samples (see Table 1). It should be noted that the DAC concentrations for maximal slope are about the same (viz., 1.9 mM DAC) for both the UM-M and the commerical samples.

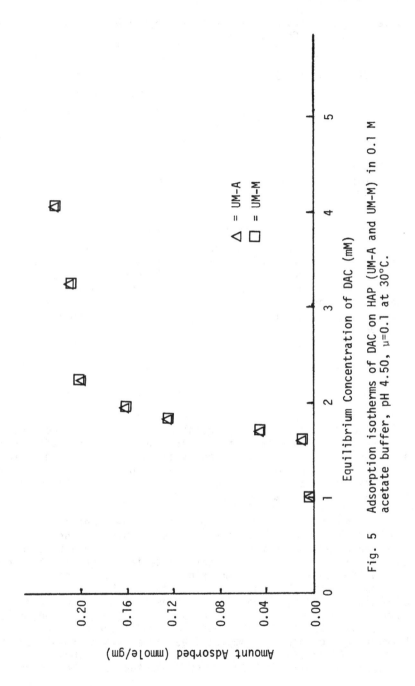

Fig. 5   Adsorption isotherms of DAC on HAP (UM-A and UM-M) in 0.1 M
acetate buffer, pH 4.50, μ=0.1 at 30°C.

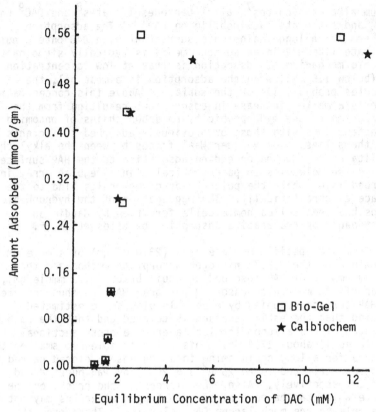

Fig. 6   Dissolution isotherms of DAC on HAP (Bio-Gel and
          Calbiochem) in 0.1 M acetate buffer, pH 4.50,
          $\mu=0.1$ at 30°C.

In order to assess the effect of the small change in pH
occurring during the adsorption experiments, isotherms at pH 4.3
and 4.8 were also obtained. They were found to be essentially the
same as those determined at pH 4.5.

DISCUSSION

The sigmoidal adsorption isotherms were common to all HAP
samples studied. This type of isotherm, although not seen as
often as the Langmuir or Freundlich types, is by no means rare
from the review of the literature. Examples are the adsorption of

sodium alkanesulfonates[7], alkyl benzenesulfonates[8] and DAC[9] on Al_2O_3 and antibiotic actinobolin on HAP[10]. The adsorption isotherm for a long chain ionic surfactant like DAC onto a polar substrate like HAP in an aqueous media is typically sigmoidal. A possible mechanism of adsorption is that at low concentration of DAC (below 1.5 mM), when the adsorption is almost nil, the molecules probably lie on the surface. Above this concentration, there is a marked increase in adsorption, resulting from the interaction of the hydrophobic hydrocarbon chains of oncoming surfactant ions with those of previously adsorbed surfactants and with themselves. The van der Waal forces between the alkyl chains facilitate the edge-on or end-on adsorption to the HAP surface, causing the molecules to pack vertically in a regular array in the adsorbed layer while the polar heads of molecules bind to the surface electrostatically. The aggregation of the hydrophobic groups has been called hemimicelle formation by Gaudin and Fuerstenau[11] or cooperative adsorption by Giles et al.[12].

From the specific surface area (23.0 m^2/gm) of sample UM-M determined by the B.E.T. nitrogen adsorption method and the maximum amount of DAC adsorbed at equilibrium (0.22 mmole/gm), the number of DAC molecules adsorbed per gram HAP and thus the area of the HAP surface occupied by each molecule can be estimated. Assuming that the entire surface is covered and that the adsorbed surfactant forms a monomolecular layer, the cross sectional area calculated is about 17.4 A^2. This value is somewhat smaller than the area for a long chain amine in a condensed, closed packed film which was estimated to be 25 A^2 and 20 A^2 by Pankhurst[13] and Adam[14], respectively. Also, the micropores and cracks on the surface which can be penetrated by nitrogen molecules may not be accessible to the much larger DAC molecules. Therefore, the actual surface area available for the DAC adsorption is probably smaller than the one obtained by the nitrogen adsorption method. This will make the calculated cross-sectional area of each adsorbed molecule even smaller. These considerations suggest that the adsorption may be multilayer, most likely approximating a bilayer which will correspond to a molecular cross-sectional area of 35 A^2 based on the nitrogen B.E.T. results.

The principal results of this study are shown in Figs. 7 and 8 where correlation between DAC adsorption and HAP dissolution kinetics are presented. At a DAC concentration of 1.9 ± 0.2 mM, it is seen that simultaneously (a) adsorption is approximately 50% of the maximum plateau adsorption, and (b) dissolution rate is about 50% of the control rate. This almost quantitative, simple inverse type relationship between inhibitor adsorption and dissolution rate, to the authors' knowledge, has not been previously reported.

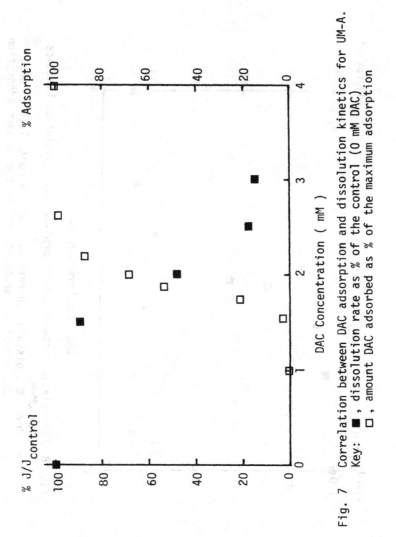

Fig. 7   Correlation between DAC adsorption and dissolution kinetics for UM-A.
Key: ■ , dissolution rate as % of the control (0 mM DAC)
      □ , amount DAC adsorbed as % of the maximum adsorption

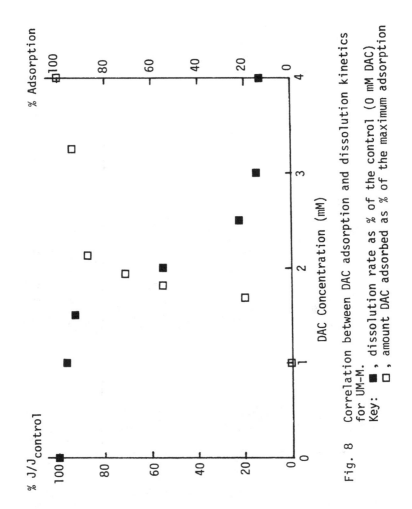

Fig. 8   Correlation between DAC adsorption and dissolution kinetics
for UM-M.
Key: ■ , dissolution rate as % of the control (0 mM DAC)
       □ , amount DAC adsorbed as % of the maximum adsorption

The following interpretation would be consistent with the observed experimental findings.

1. Dissolution rate is essentially zero where there is adsorption of DAC. This may be reasonable if sigmoidal nature of adsorption isotherm, as discussed before, is interpreted as one in which there is considerable cooperativity among the DAC molecules in the adsorbed state leading to patchwise adsorption of "islands" through which ionic diffusion is extremely difficult. Such closely packed islands might be expected to completely stop the dissolution from the HAP surface beneath them. The observed dissolution rate may be therefore directly related to the area of uncovered surface. This mechanism would differ from the kink poison mechanism proposed by Napper and Smythe[15]. They attributed the decreased dissolution rate of HAP in the presence of organic and inorganic phosphate to kink stabilization or kink poisoning whereby adsorption at kink sites on the crystal surface decreased the ion disengagement rate. The kink site adsorption and the maximum rate reduction would be expected to occur at very low surface coverages because the high energy kink sites were thermodynamically preferred for adsorption. The kink site poisoning mechanism would fail to explain the observed effect of DAC upon the HAP dissolution. The sigmoidal nature of the isotherm also precludes this mechanism since additive adsorption at kink sites would be expected to be closer to Langmuirian as Gallily and Friedlander[16] have suggested.

2. The dissolution kinetics is surface controlled. This was shown previously[2] for HAP dissolution under the same conditions but in the absence of DAC. If the dissolution is diffusion controlled, the rate would be expected to remain relatively uninfluenced by adsorption until most of the surface is covered; this would be inconsistent with the present results.

3. If the predominant surfaces (for example, the prism planes) of HAP crystal were involved in DAC adsorption, then the same surfaces would be involved in the surface controlled dissolution kinetics. Because the component of the dissolution rate in the control situation[2] was previously designated as involving site #2, we propose to designate site #2 as being associated with predominant surface(s) of the HAP crystal.

There can be several interpretations for the residual dissolution which is about 15% of the maximal rate and is unaffected by increasing the DAC in the solution. Although maximum adsorption appears to take place before the CMC is reached, it may be arqued that this may not necessarily be the case because of the experimental uncertainties involved. This adsorption on the remaining unadsorbed surface does not take place because, beyond the CMC, the DAC monomer activity may increase

only very little with increasing solution DAC concentration.
Another possibility is that the coverage may be essentially
complete but there are micropores in the surface which are not
accessible to the large DAC molecules due to, for example, steric
factors. This uncovered area may contribute to the limiting
dissolution rate. A third possible mechanism is that there exists
two or more distinct types of sites on the crystal: one site
which is inhibited by DAC and which comprises most of the area of
the crystal, and other site(s) not affected (or much less
affected) by DAC and which makes up about a 15% contribution to
the total rate of dissolution of the crystal under sink condition.

Table 3. The Surface Area and the Corresponding Dissolution Rate
          of Different HAP Samples in 0.1 M Acetate Buffer,
          pH 4.50, μ = 0.1

HAP	S.A.[a]	S.A./S.A.$_{Y155}$	J_0[b]	$J_0/J_{0,Y155}$
NBS-P	22.5	9.78	1.57	12.76
UM-A	23.4	10.17	1.30	10.57
UM-M	23.0	10.00	1.60	13.00
UM-K	22.9	9.95	2.00	16.26
Y155	2.3	1.00	0.123	1.00

[a] S.A. is specific surface area in m^2/gm.
[b] J_0 is dissolution rate of control in ppm PO_4/sec.

Comparisons of the results obtained from the different
samples in this study reveal interesting similarities and
differences. Sample Y155, a high temperature preparation, had the
slowest dissolution rate. Its control rate (without DAC) was
about 10% of other samples. However, the shape of the J vs DAC
concentration curve (Fig. 4) and the DAC concentration at 50% rate
reduction for Y155 are similar to those for sample UM-M. The
above observations support the interpretation that hydrothermally
prepared HAP may differ from HAP prepared by 100°C precipitation
and digestion only with regard to surface area (see Table 3) and
that its dissolution rate behavior may be otherwise the same.
Interestingly, samples with 0.3% carbonate (UM-T) and no carbonate
detectable by IR behave virtually identically, indicating that the
presence of carbonate does not measurably alter the effect of
DAC. Another sample (Y146) with a relatively high HPO_4 content
had a much more gradual dependence of dissolution rate on the
concentration of DAC. That the HAP dissolution inhibition by DAC

may be altered by HPO_4 to such degree might have clinical implications and needs further study.

REFERENCES

1. J. L. Fox, W. I. Higuchi, M. B. Fawzi and M. S. Wu, A new two-site model for hydroxyapatite dissolution in acidic media, J. Colloid Interface Sci. 67:312 (1978).
2. W. I. Higuchi, E. Y. Cesar, P. W. Cho, and J. L. Fox, Powder suspension method for critically re-examining the two-site model for hydroxyapatite dissolution kinetics, J. Pharm. Sci. accepted for publication.
3. T. J. Roseman, W. I. Higuchi, B. Hodes, and J. J. Hefferren, The retardation of enamel dissolution rates by adsorbed long-chain ammonium chloride, J. Dent. Res. 48:509 (1969).
4. Y. Avnimelech, E. C. Moreno, and W. E. Brown, Solubility and surface properties of finely divided hydroxyapatite, J. Res. Nat. Bur. Stand. (U.S.) 77A, 1:149 (1973).
5. A. Gee, L. P. Domingues, and V.R. Deitz, Determination of inorganic constituents in sucrose solutions, Anal. Chem. 26:1487 (1954).
6. T. Ino, T. Kondo, and K. Meguro, J. Chem. Soc. Japan, Pure Chem. Sec., 76:220 (1955).
7. T. Wakamatsu and D. W. Fuerstenau, The effect of hydrocarbon chain length on the adsorption of sulfonates at the solid/water interface, in: "Adsorption From Aqueous Solution," W. J. Weber, Jr. and E. Matijevic, ed., American Chemical Society, Washington D.C. (1968).
8. S. G. Dick, D. W. Fuerstenau, and T. W. Healy, Adsorption of of alkylbenzene sulfonate (A.B.S.) surfactants at the alumina - water interface, J. Colloid Interface Sci. 37:595 (1971).
9. B. Tamamushi and K. Tamaki, Adsorption of long-chain electrolytes at the solid/liquid interface, Trans. Faraday Soc. 55:1007 (1959).
10. D. E. Hunt and J. K. Hunt, Adsorption of the antibiotic actinobolin by hydroxyapatite, Archs. Oral Biol., 25:431 (1980).
11. A. M. Gaudin and D. W. Fuerstenau, Quartz flotation with cationic collectors, Trans. AIME, 202:958 (1955).
12. C. H. Giles, A. P. D'Silva, and I. A. Easton, A general treatment and classification of the solute adsorption isotherm, J. Colloid Interface Sci., 47:766 (1974).
13. K. G. A. Pankhurst, The adsorption of paraffin-salts to proteins, Faraday Soc. Discussion, 6:52 (1949).
14. N. K. Adam, "The Physics and Chemistry of Surfaces," Oxford University Press, London (1941).

15. D. H. Napper and B. M. Smythe, The dissolution kinetics of hydroxyapatite in the presence of kink poisons, J. Dent. Res. 45:1775 (1966).
16. I. Gallily and S. K. Friedlander, Kink poisons and the reduction of dissolution rate at crystal-liquid interfaces, J. Chem. Phys., 42:1503 (1965).

SURFACE DEPENDENT EMISSION OF LOW ENERGY ELECTRONS (EXOEMISSION)

FROM APATITE SAMPLES

J.E. Davies

Department of Anatomy
University of Birmingham
Birmingham B15 2TJ. UK.

ABSTRACT

Exoemission (EE) phenomena, grouped collectively under the
title "Kramer effect" include non-isothermally stimulated relaxation
of irradiation-induced perturbations of crystalline lattices known
as Thermally Stimulated Exoemission (TSEE). Liberated electrons
may originate from metastable excited states via the conduction band
(volume effect) or from phase changes of surface adsorption species
(surface effect). However, in all cases, changes in the surface
state may modify the emission. This paper presents TSEE observed in
apatites using an open-window Geiger counter. In a "model" series,
emission maxima common to FAp's of synthetic mineral and biological
origins demonstrate surface and volume dependent emission within the
series. Similar emission characteristics observed from biological
apatites of differing origins indicates that the defects responsible are intrinsic to the apatite lattice and at least partially
independent of chemical substitutions. Since electron availability
is of fundamental importance to biological hard tissue reactivity,
the potential of exoemission as an adjunct to other surface
analysis techniques is discussed.

FOREWORD

The biological hard tissues, bone, dentine and enamel, comprise
organic and inorganic fractions. The major inorganic component
common to these tissues is a highly mineralized, chemically complex
form of calcium hydroxyapatite[1]. Physicochemical characterization of
this constituent of hard tissues is therefore important to an overall
understanding of its biological function.

71

Once fully formed, dental enamel is a dead tissue, changed
only by insults from its external environment; perhaps the most
obvious examples being pathological conditions such as dental caries
and the mechanical insult comprising the preparation of teeth for
restorative procedures. In each of these cases the remaining enamel
apatite (96-98% by weight) will be changed in some respect.
Crystalline lattice defects are known to play an important role in
diffusion and dissolution processes that occur in enamel[2] while
secondary electron flaring has been shown to occur on mechanical
deformation of this tissue[8].

Bone, in contrast, is a dynamic, living tissue exhibiting a
quantifiable turnover rate which changes, but nevertheless continues,
throughout life. Bone apatite activity, resorption and deposition,
will involve ongoing surface exchange of charged particles, the
nature of which will influence not only the rate but also the type
of ionic activity associated with this tissue surface. The
biological reactivity of bone therefore, while being influenced by
volume crystalline lattice defects, will depend predominantly upon
the apatite/collagen or apatite/tissue fluid interface.

Although enamel and bone apatite crystallite dimensions are
known to be significantly different, lattice defects will in both
cases occur as a result of deviations from stoichiometry and may be
common to apatites of different origins.

Work reported in this paper shows that such defects, intrinsic
to the common apatite lattice, may be demonstrated using a physico-
chemical phenomenon known as "Exoemission".

INTRODUCTION

Deviations from stoichiometry in the crystalline lattices of
insulating materials such as apatite will result in the availability
of energy levels, localised within the energy band gap, at which
electrons and holes may become trapped, creating a thermodynamic
non-equilibrium or metastable state. Relaxation of these created
perturbations will restore the equilibrium of the system but is
dependent upon an external energy supply. This phenomenon is the
basis upon which many solid-state experimental observations are made.

Optical absorption (OA) and photo conductivity (PC) are methods
employing photon energy as the external energy supply and have been
used for the investigation of colour centres in apatites of synthetic
(Table 1) and biological[9,10,11] origins, while thermoluminescence (TL)
and thermally-stimulated conductivity (TSC) are examples of a group
of phenomena which require a temperature reservoir to activate the
relaxation process.

Table 1. Comparison of published values of Optical Absorption (OA), Photoconductivity (PC) and Optically Stimulated Exoemission (OSEE) for synthetic fluorapatites.

METHOD	SAMPLE	REF.	TEMP.	WAVELENGTH NM				
OA	SC $\parallel$ c-axis	(3)	RT	750	610		450	
R	POWDER	(4)	RT		610		450	370
OA	SC $\parallel$ c-axis	(5)	77K	735	600		440	
"	SC $\perp$ c-axis	"	77K					350
PC	SC $\perp$ c-axis	"	200K					380 350
OSEE	POWDER FApI	(6)	RT				450	370
OSEE	POWDER FApII	"	RT			520	450	370

Manganese absorption bands in FAp 160, 175 and 215nm (7)

Thermally stimulated relaxation processes (TSR's) have recently been extensively reviewed by both Braunlich[12] and Chen and Kirsh[13]. Of this group, the thermoluminescent properties of apatites have been investigated from differing points of view. Maletskos[14], Kastner et al[15], Bordell and Kastner[16] and Willhoit and Poland[17] investigated the possibility of using TL for clinical irradiation dose measurements using bone mineral. Jaskinska and Niewiadomski[18], Christodoulides[19,20], Christodoulides and Fremlin[21], Driver[22] and Benko and Koszorus[23] measured TL in bones and teeth and Wintle[24] and Bailiff[25] in apatite mineral grains for archaeological dating purposes; while geological dating of apatites by TL was discussed by Nambi[26], Vaz[27] and Alves et al[28]. TL was reported while investigating mixed chlor-fluorapatites for their use in the phosphor industry[29,30] and Lapraz et al[31] studied the role of various activators in TL from both synthetic and mineral hydroxyapatite. Kolberg et al[32] and Chapman et al[33] have used TL as a means of investigating the fundamental properties of biological hard tissue and a means of comparing them with chemically simpler apatites.

Many of the above authors reported difficulties in the measurement of TL in apatites and in particular those who examined biological material. Full discussion of these problems is outside the scope of this paper but the most consistently reported complications were, firstly, that biological material exhibited chemiluminescence which in some cases masked volume emission and secondly, that thermal pretreatment to high temperatures was often necessary to render the material thermoluminescent. Some concluded[34] that in biological material surface light emission dominated and therefore prevented recording of volume effects. However, in the biological environment and in particular bone apatite, surface properties will

be of considerable importance and these TSR phenomena therefore
warrant reappraisal.

EXOEMISSION

The term "exoemission" describes a group of emission phenomena
which occur during relaxation of perturbations of thermodynamic
equilibria in the volume or at the surface of solids. These
perturbations may be as a result of mechanical, optical, chemical
or high-energy radiation treatment of the solid.

The structure-dependence of the effect was discussed in 1953[53]
and the term "Kramer effect", although now rarely used, was adopted
due to the latters' early systematic work[35,37] in this field.
Analogies and correlations with TL and TSC led to kinetic theories
based on those established in TL. Numerous review articles have
been published in the proceedings of international symposia on
exoemission together with that recently published by Glaefeke[38].

Emission may occur spontaneously, during or following a
perturbation, but the two cases referred to in this paper require
an energy source to allow return to equilibrium during which an
emission current is measured. Energy supplied as photons will
result in optically stimulated exoemission (OSEE) or from a temperature ramp, thermally stimulated exoemission (TSEE).

OSEE may be considered as a special case of the external photoeffect[39] and may exhibit structured emission with peaks corresponding
to optical absorption bands (selective-OSEE) or direct optical
ionization with emission increasing with the stimulating energy
(non-selective OSEE)[40]. Selective OSEE is considered to be a two-
step photo-thermal ionization process where, following an optically
excited transition to the bottom of the conduction band, electrons
escape from the latter by a thermionic process (see equation (i)
below and refs 38,40 for review of emission models). The plot of
optically stimulated electron emission against wavelength is called
an OSEE spectrum. As the current density is temperature dependent,
increasing T above ambient in OSEE will increase the probability of
escape up to a temperature maximum corresponding to that required to
thermally anneal the perturbations.

TSEE may also be explained in some instances as being directly
related to volume defects[41] but here, interpretation is complicated
by the fact that exoemission may occur directly from intrinsic or
adsorption induced surface states[42]. The plot of thermally stimulated electron emission against temperature is called a TSEE glow
curve. Thus, for thermally stimulated exoemission, two broad
theoretical concepts have been established:-

(a) THE VOLUME CONCEPT is based on the assumption that

electrons may escape from thin surface layers. Volume
solid-state reactions are involved in the delocalization
of charge-carriers, electrons and holes. Electrons
providing the output current are emitted via the conduction band. A thermionic model is most often used to
explain volume dependent emission where the emission
current from the conduction band will be[38]:-

$$J_o = n_c (\frac{kT}{2m})^{1/2} e^{-\chi/kT} \quad \ldots \ldots \ldots \ldots \ldots \ldots \ldots \text{(i)}$$

where n_c = concentration of electrons at the conduction
band edge, K= Boltzmann's constant and m = electron
mass. Surface states will influence the probability of
escape and the energy distribution[43].

(b) THE SURFACE CONCEPT is based on the following criteria[44]:-

   i. surface layer destruction as a result of mechanical,
       radiation or other factors resulting in the format-
       ion of active particles and subsequent desorption
       of gaseous products.

   ii. active particle recombination in the surface layer
       providing energy for the emission of electrons.

   iii. escape of electrons, without transition to the
        conduction band.

This may be summarized using the following simple equations[38]

   excitation : $A_2 + e^- \rightarrow A_2^-$,

   stimulation : $A_2^- + E_e \rightarrow A_2 + e^{-\dagger}$ $\ldots \ldots$ (ii)

Where A = atom A, e^- = electron, E_e = activation energy for
TSEE (eV) and $e^{-\dagger}$ = exoelectron.

However, as a three-dimensional surface will represent a
transition zone from the ordered crystalline lattice to the vacuum
it is inevitable that both volume and surface phenomena are
intimately associated and possibly interactive.

MATERIALS AND METHODS

Although no synthetic apatite can be regarded as a realistic
model of an apatite of biological origin, due to the chemical
complexity of the latter, synthetic fluorapatites (FAp) were nevertheless chosen as a starting material in the present work for three
reasons. First, considerable physical information is available

(see for example:- 3,4,5,7, and references therein) concerning
synthetic fluorapatites due to the investigation of lamp phosphors.
Secondly, data is available for comparison so that one can assess
the validity of a relatively new technique such as exoemission.
Third, fluorapatites occur both in mineral deposits and certain
biological materials.

Sample batches of two powder synthetic FAp's (named FAp.I and
FAP.II here: ref 6), one mineral FAp (Powdered crystals of Durango
FAp) and powder prepared from the surface of the enameloid of a
shark tooth (Great White) were used in this investigation as a
synthetic-mineral-biological FAp series. Comparison with a biological hydroxyapatite was afforded by using enamel prepared from a
human premolar. Synthetic FAp.I was prepared by G.E.C. Research
Labs although the details of preparation are not known to the
author. Synthetic FAp.II was prepared by heating 7.8g CaF_2 and
93g β-tricalcium phosphate for a total of 7 hours at $1000^{o}C$. Shark
enameloid was chosen as an end-member of this synthetic-mineral-
biological FAp series as the surface of marine shark enameloid is
known to correspond to nearly saturated fluorapatite[45]. This
powder was obtained by carefully scraping the dry surface of the
tooth with a glass microscope slide to remove the "shiny" surface
only. No chemical extraction was employed.

The human dental enamel powder was prepared from a caries-free
premolar which had been extracted for orthodontic reasons. The
root was separated from the crown, coronal dentine and the bulk of
enamel removed from the pulpal side and discarded: the remaining
thin enamel cap was broken into fragments and pulverized using an
agate mortar and pestle.

Powders (grain size < 100 μm) were mixed with analar acetone
and shaken. After initial settling to remove the larger grains the
supernatant suspension was removed with a pipette and sedimented, by
evaporation at room temperature, onto stainless steel disks of 7mm
diameter and 1mm thickness. A series of scanning electron micrographs of an FAp.I preparation is shown in Figure 1. Individual
grains are typically 9-15 μm and generally of square platelet form.
They are composed of fused hexagonal plates allowing containment of
spaces which themselves exhibit a roughly hexagonal outline. Both
synthetic materials showed similar morphology. Due to the thickness
of both the stainless steel supporting disk and the powder layer
itself, temperature values on the X-axes of all figures refer to
sample base temperature (the thermocouple is sited immediately
beneath, and in contact with, the sample disk) and not to the sample
surface temperature.

Figure 2 shows a block diagram of the apparatus used here for
OSEE, TSEE and TL measurements. The electron counting chamber, an
open-windowed Geiger-Muller type, heater controller and stove

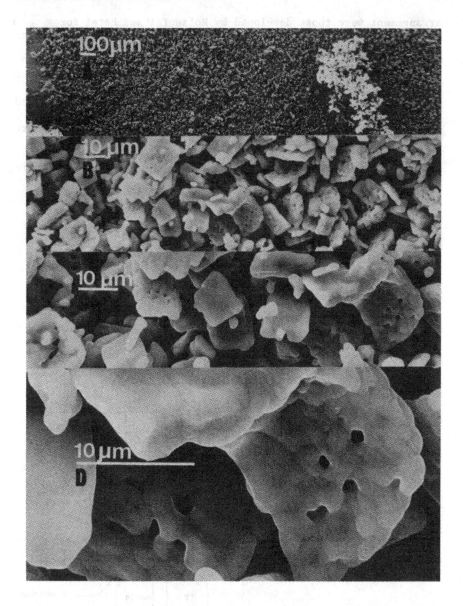

Figure 1. Scanning electron micrographs of FAp.I, sedimented onto
a stainless steel disc as described. At high magnif-
ication note hexagonal platelet form.

arrangement were those developed by Holzapfel and Petel for
exoemission dosimetry investigations and have been fully described
by them elsewhere[46,47]. The chamber is furnished with a quartz
window allowing either entry of stimulating light, from a 150 W
quartz halogen lamp for OSEE, or the detection of emitted light via
a U-V fibre optic to a photomultiplier tube (EMI 9804 QB) for TL
measurements. A large filter carrying disc (16 filters) is placed
in the light beam of the OSEE facility for spectral measurements.
Background counts at ambient temperature for electron counting are
approximately 1 count.sec^{-1} but as high as 1200 counts.sec^{-1} for
photon detection since the PMT is not cooled. Both electron and
photon counting are processed using standard Ortec nuclear counting

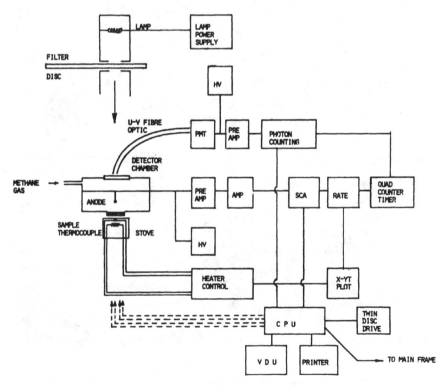

Figure 2. Block diagram of apparatus for measurements of OSEE,
           TSEE, TL, simultaneous OS/TSEE and simultaneous TL/
           TSEE. Dotted lines represent CPU control of stove
           as an alternative to "HEATER CONTROL".

equipment (E.G. & G. Brookdeal Electronics, Berkshire U.K.) and
signals are recorded using either an X-Yt plotter or, more recently,
logged using a Research Machines 380 Z microprocessor as a central
processing unit (CPU). This apparatus, together with a CPU
controlled stove arrangement is described fully elsewhere[48].

Samples were excited using a 20mCi β^{90}Sr source. Irradiation
times or doses are indicated where appropriate. Thermal scanning was
carried out, in all measurements reported here, at $1^{\circ}C.sec^{-1}$ from
ambient temperature to $600^{\circ}C$. In all sequential measurements on one
sample, cooling conditions, irradiation times and delay time between
readings were standardized unless otherwise stated in the text.

RESULTS

OSEE-Synthetic Fluorapatites

Synthetic fluorapatites were chosen for the first stage in
examining the exoemission behaviour of apatites due to the availability of data concerning their optical absorption properties.
Since congruity of OA and OSEE had been elegantly demonstrated by
Nink and Holzapfel, in the simple cubic lattice of $CsCl^{49}$, OSEE
spectra were obtained for two "stoichiometric" synthetic FAp
preparations.

Distinct similarities were found[6] in the values of the OSEE
and OA peaks (see Figure 3 and Table 1) the major emission occurring
at 370 and 450 nm. Furthermore, not only did monochromatic optical
stimulation of either emission peak cause partial reduction of the
other, demonstrating a relationship between the electron trapping
mechanisms for both, but also, using rather simple calculations, the
F^- ion chains were implicated in this defect structure. This was
considered as strong evidence for the volume dependence of optically
stimulated exoemission in these samples.

However, a reproducible difference between these "stoichiometric" preparations could be found, manifesting itself as a longer
wavelength shoulder (lower stimulating photon energy) on the 450 nm
OSEE peak, associated with samples thermally pretreated at $950^{\circ}C$ in
air. Selective OSEE sensitivity appeared specific to an FAp
group of synthetic and mineral origin while ClAp's, HAp's and BrAp
exhibited only non-selective OSEE (unpublished data).

TSEE-Synthetic Fluorapatites

Using the same synthetic FAp samples, TSEE, although a less
sensitive emission process (less emission counts per unit of
excitation energy), also demonstrated two dominant peak structures
with a low temperature shoulder on the first peak in those samples

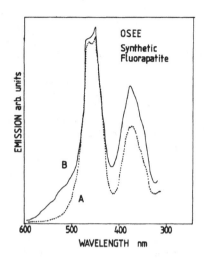

Figure 3.   OSEE spectra of synthetic FAp.I before (A) and after (B)
            thermal pretreatment in air at 950°C for 2 hours —
            redrawn from ref. 6.

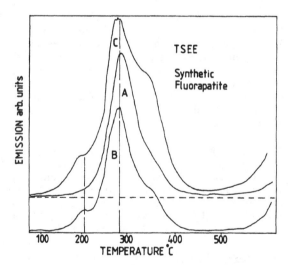

Figure 4.   TSEE.   Synthetic FAp's.I. and II (curves A and B) and
            synthetic FAp.I preheated in air for 15 hours at 1000°C
            (curve C).   Compare Figure 3.

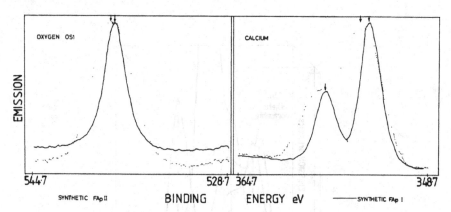

Figure 5. ESCA analysis. FAp.I. and FAp.II. showing distributions of binding energies for calcium and oxygen l.s. It is the distribution form of the curves and not the labelled energy peak values which are being compared here.

which had been thermally pretreated as above (see Figure 4). It is assumed that this thermal pretreatment caused fluoride/oxygen ion substitutions in the host lattice and hence changed the TSEE glow curve strucutre.

Interestingly, preliminary ESCA (electron spectroscopy for chemical analysis) measurements have shown that significant differences can be seen between these samples in the distribution of their electron binding energies of calcium and oxygen (see Figure 5) and an empirical evaluation of these results would lend support to the hypothesis that surface exchanged oxygen ions are responsible for the differences in both OSEE spectra and TSEE and glow curves respectively.

TSEE glow curves in synthetic FAp's, although consistently exhibiting structures at the same temperatures, seemed to be of unreproducible sensitivity both in total emission counts and the heights of the peaks relative to each other and, therefore, purely volume concepts cannot be applied to TSEE in these materials. These differences are clearly shown in Figure 6. During the first thermal scan, triboemission is seen at high temperature (thermal relaxation of mechanically excited states induced during sample preparation) in addition to irradiation induced peaks at 275°C and 350°C. The increase in emission at 350°C seen in curve (c) Figure 6, recorded following irradiation, after having left the sample exposed to the laboratory atmosphere overnight would suggest that emission at this

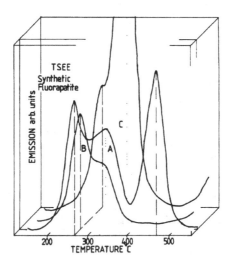

Figure 6.   TSEE synthetic FAp.I. each recording after 2 mins
             irradiation β[90]Sr. A. 1st heating cycle. Note
             triboemission at high temperature. B. 2nd heating.
             No triboemission. C. 3rd heating. Irradiated and
             recorded after having left the sample in air overnight
             following (B).

temperature is strongly dependent upon surface states. With
repeated irradiations and thermal cycling, the height of this peak
is reduced until it becomes a shoulder on the high temperature tail
of the 275°C peak (Figures 4,7). However, it should be noted that
further periods of exposure in air before re-irradiation can show
a general decrease in emission of the whole glow-curve including the
275°C peak although the first re-reading after such a delay always
demonstrates the peak at 350°C at a higher level than that of 275°C.

     Due to the exothermic connotations of both the original term
"exoelectrons" and the more recent physico-chemical surface reaction
emission concepts of Krylova[50], attempts were made to monitor exo/
endothermic reactions during an on-going thermal scan of this synthetic FAp using both DTA (Differential Thermal Analysis) and DSC
(Differential Scanning Calorimetry). No significant deviation from
the base line could be established with either technique and the
broad endotherms associated with some mineral FAp's[51] could not be
identified in these synthetic samples. This is not the case with
apatites of biological origin. Holcomb and Young demonstrated peaks

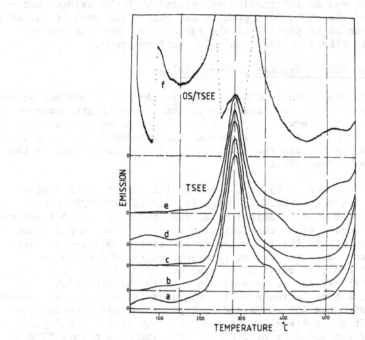

Figure 7. FAp.I. 2 mins irradiation prior to each glow curve recording. Curves (A),(B) and (D) TSEE. Curves (C) and (E) TSEE
after optical bleaching. Curve (F) combined OS/TSEE.

in human dental enamel at approximately 100°C and 400°C, using
differential thermogravimetric analysis (DTGA), corroborating
earlier work[52]. Changes in the 400°C region were assigned to
decomposition of both organic and inorganic components with
concomitant changes in the apatite lattice parameters. The TSEE
signals in both human dental enamel and shark enameloid in the
temperature ranges 200°C-500°C and 300°C-500°C respectively, seen
in the 1st heating cycle and shown in Figure 10.1 curve (a) and
Figure 11.1 curve (a) could be associated with surface mediated
reactions related to such chemical changes. The fact that thermal
analysis provided no information associated with the evident surface
dependence of the 350°C TSEE peak is not surprising when it is noted
that the exoemissive portion of the surface is, according to
Krylova[44], in alkali halides, only 10^8 surface atoms.

Combined OSEE/TSEE - Synthetic Fluorapatites

It would appear that OSEE and TSEE in FAp are volume and surface
related respectively. Optical stimulation at wavelengths corresponding to OSEE peaks (known as optical bleaching) prior to measurement
of the TSEE response and simultaneous measurements of OSEE and TSEE
from the same sample can therefore give information concerning the
relationship between the two.

Figure 7 curves (a-f) show a sequential series of TSEE glow
curves from synthetic FAp.I. Curves (a) and (d) are TSEE glows
immediately following irradiation (approx. 600 rads) without optical
bleaching. Curve (b) is the TSEE glow from the same sample recorded
after a delay of 45 mins following irradiation. Emission has been
partially reduced in the region from ambient to 50°C while the
remainder of the glow curve is unchanged. Curve (c) was recorded,
following irradiation, after a 45 min. optical bleach using a
Balzers B40, 446 nm narrow band interference filter (this was chosen
to compare with the 450 nm OA and OSEE peaks in this material (see
Table 1). It shows bleaching of the TSEE glow to 150°C and partial
bleaching of the shoulder at 350°C. Recording of a normal TSEE glow
curve (d) demonstrates that both the low temperature emission and the
shoulder at 350°C are re-instated but there is also an emission
increase in the temperature region 480-500°C. A further 446 nm
bleach prior to recording the TSEE again reduces both the low
temperature emission and the shoulder at 350°C but the high temperature emission is becoming increasingly evident (curve (e)). It
would seem therefore, that optical bleaching of the TSEE response
using a wavelength corresponding to one of the absorption bands of
FAp has little effect on the 275°C peak. However, the 350°C shoulder
is bleached and appearance of the high temperature emission may
possibly be due to phototransfer.

Curve (f) of Figure 7 shows the combined OS/TSEE in the same
sample following the same irradiation dose using the 446 nm filter

and is characterised by the following:-

  i) the OS/TSEE emission sensitivity is greater than individual
      measurements of OSEE or TSEE (up to 450°C)

 ii) there is a distinct valley in the glow curve at approximately
      150°C

iii) an emission peak at 275°C corresponding to that seen in TSEE

 iv) above 275°C the glow curve corresponds to that in TSEE from
      samples which have been previously optically bleached (446 nm
      filter) i.e. there is no shoulder at 350°C but high temperature
      emission at 480°-500°C is evident

  v) no significant difference in the high temperature emission for
      the same irradiation dose.

     The height of 500°C emission, unlike that of 275°C is unaffected
by combined OS/TSEE although the rise in thermionic emission towards
the temperature maximum would seem to be decreased.

     High temperature TL emission in the 480°-500°C range in these
samples was discussed in a recent paper[28] and compared with the
"natural" TL found in Durango FAp. This emission peak has not been
seen in TSEE glow curves alone (whereas using a 348 nm filter it can
be seen routinely in TL measurements) and is only apparent after
prior optical bleaching of the sample or simultaneous OS/TSEE
measurements.

     OSEE in this case is completely bleached simultaneous with the
drop in exoemission on the high temperature tail of the 275°C peak.

     As OSEE is so closely associated with OA, OS/TSEE may be
compared with TBOA (Thermally Bleached Optical Absorption) which has
been shown in alkali halides[40] to occur at temperatures corresponding
to annealing of TSEE emission. Samuelsson[43] demonstrated similar
behaviour in LiF and assumed that the high temperature glow peak in
LiF is associated with thermal destruction of F-centres. It would
seem reasonable that the congruent thermal bleaching of OSEE with
the 275°C TSEE peak is evidence of volume traps being involved in
TSEE from FAp although it must be stressed that this does not allow
assignment of the 275°C peak to specific defects.

TL/TSEE - Synthetic Fluorapatites

     Comparison of TL and TSEE glow curves (see Figure 8) shows
superimposition of emission at 275°C affirming the volume dependence
of emission at this temperature. However, 3D plots of TL intensity

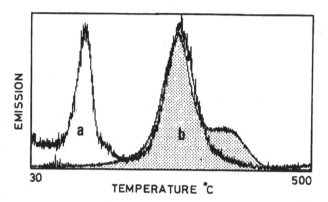

Figure 8.   TL (a) and TSEE (b) synthetic FAp.I.   Emissions are
congruent at 275°C.

and spectral dependence (Figure 9) show that light is emitted at
this temperature at two different wavelengths (400 and 570 nm).
Ratnam et al[30] ascribed the occurrence of an emission peak in
synthetic fluorapatite at 570 nm to Mn^{2+} ion   substitutions for
Ca^{2+} which agreed with 570 nm fluorescence emission reported by
Ryan et al[52] while emission at 420 nm, ascribed to Sb^{3+} ions did not
occur in their samples above 200°C.   The latter is evidently not
the case with the FAp.I. examined here.

The shift from blue to green emission at 275°C seen after the
first heating cycle suggests that diffusion of TL activator sites
has occurred during, or after, the first thermal scan.   Diffusion
of activators, in particular Mn^{2+} outside the lattice of hydroxyapatite was discussed by Lapraz et al[31] as a possible mechanism
responsible for simultaneous decreasing of all TL peaks observed
after thermal annealing, although, using electronic microprobe
measurements they found no evidence of Mn^{2+} shift after annealing
Mn-doped chlorapatite.   If it is assumed that the 400 nm emission
seen here is due to Sb^{3+} ions, then their diffusion, during an
ongoing non-isothermal scan, would be facilitated, with respect
to Mn^{2+} ions, due to their smaller ionic radius (0.76Å compared
with 0.80Å).   This would account for the decrease in emission of
light at this wavelength while migration of Mn^{2+} ions within the
lattice may allow an increase in emission of light at 570 nm.
Ryan et al[53] showed conclusively that Mn^{2+} substitution at Ca[I]
(on the threefold rotation axes parallel to the c-axis) was
predominantly associated with fluorescence even in preparations
containing a Mn[II]:Mn[I] ratio of 1800:1.   Preliminary flame

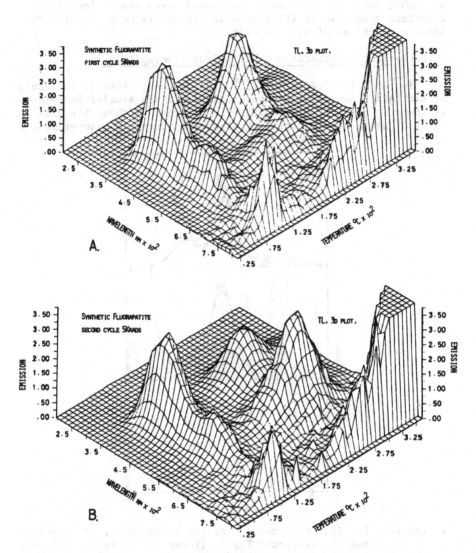

Figure 9.   3D plots : TL temperature and spectral dependence of
FAp.I. (a) 1st heating cycle (b) 2nd heating cycle.
Note spectral shift in 270–280°C region.

absorption spectroscopy measurements show that the manganese level
in FAI (80-88 ppm) drops to approximately 60 ppm on heating in air
at 1000°C for 15 hours. Use of extended X-ray absorption fine
structure spectroscopy (EXAFS) could, in future work, clarify the
lattice positions of such impurities.

TSEE – Synthetic, Mineral and Biological Fluorapatites

Increase in TSEE in the 275°C region can be seen in the FAp's
of synthetic, mineral and biological origin and similar behaviour
is found in their TL glow curves[54]. Figure 10.1 shows the glow
curve of synthetic FAp (refer to Figure 8, curve (b)) superimposed

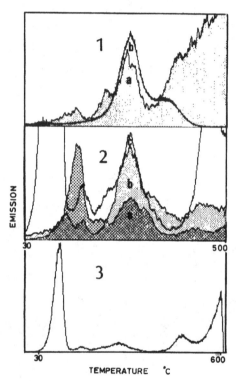

Figure 10.   1)   TSEE glow curves (a) shark enameloid, 1st heating,
             and (b) synthetic FAp.I. (refer to figure 8 curve (b)).
             2)   TSEE glow curves (a) shark enameloid (b) synthetic
             FAp.I. (previously used for optical bleaching measure-
             ments) (c) Durango FAp - Y axis expanded to show mid-
             temperature region.
             3)   TSEE Durango FAp original curve used in 2(c) above.
             Note change of X-axis scale.

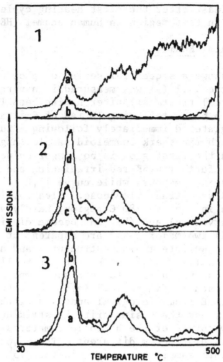

Figure 11.   TSEE glow curves : Human enamel. Sequential series of
             heat cycles 1(a) and (b): 2(c) and (d). 3(a) : - curve
             2(d) above compared with curve 2(a) from Figure 10.

on that of irradiated shark enameloid. The powdered enameloid had
undergone no chemical or thermal pre-treatment and emission at high
temperature is probably associated with both tribo- and chemi-
emission. Nevertheless, a complex structured emission is evident
up to 350°C with a double peak centred at 275°C. Figure 10.2
compares the emission of synthetic FAp, Durango FAp and shark
enameloid. In this figure the synthetic material had previously
been used for optical bleaching experiments and therefore the high
temperature emission is clearly seen, while the shark enameloid
glow curve represents the fourth heating cycle with this sample and
triboemission is no longer evident. The scale of the Durango FAp
glow curve has been expanded to visualize the 275°C peak; the
original glow curve is shown in Figure 10.3.

     The common occurrence of TSEE at this temperature in all FAp's
examined was considered previously[55] to indicate a volume phenomenon
solely characteristic of the FAp series. However, with the increased
data logging facilities of the CPU and the resultant improvement in
observing low sensitivity emitters such as the HAp's, it can be

shown that, at least after the first heating cycle, emission is clearly present in this region in human enamel (HE).

TSEE - Human Dental Enamel

Figure 11 shows a sequential series of glow curves observed in HE. Glow curve 11.1 (a) was measured in a virgin sample (i.e. no previous thermal treatment) after having been left at room temperature for 5 hours following 1M rad of β-irradiation. Curve 11.1 (b) was registered immediately following a further dose of 1.5K rad. As with the shark enameloid sample, high temperature emission seen in the first glow is no longer evident. Curve 11.2 (c) represents a further 4.5K rad irradiation and a resultant increase in emission response while curve 11.2 (d) was recorded following a 273K rad irradiation dose. Clearly, in these two glow curves, emission in the 240-280°C region, hardly visible above background in curves (a) and (b), has become distinctly structured. In curve 11.2 (d) two major peaks are apparent at 130°C and 270°C with evidence of two intermediate structures and a third on the high temperature tail of 275°C peak. In figure 11.3, curve (a) is the same as that shown in 11.2 (d) but superimposed on the shark enameloid curve used in figure 10.2 (a). These two curves there-fore represent TSEE from biological apatites of the HAp and FAp type respectively and their similarity is striking. Change in form of the glow curves of the biological materials since the first heating cycle, is distinctly different. The major initial structure in the virgin shark sample being seen in the 275°C region while that of HE is at 130°C. With repeated cycling the 130°C peak becomes dominant in both samples.

CONCLUSIONS

Table 2. summarizes the principle TSEE peaks seen in the synthetic, mineral and biological fluorapatites investigated. While only TL is seen in the synthetic samples in the 100°C region, indicating a purely volume phenomenon, the TL, TSEE and combined OSEE/TSEE observed in the temperature range from 210°C to 350°C demonstrates an interaction between volume and surface effects. Congruence of emission in all FAp's and HE in this mid-temperature region, in spite of considerable differences in sensitivity, indicates that not only is it the apatitic, rather than the organic component of the biological materials which is responsible for the emission but that the defects responsible are common to apatites of different origins.

The increasing irradiation dose/emission response and resolut-ion in the 275°C region, see in figure 11, probably represents diffusion of defects to the sample surface. While defect diffusion processes occurring during thermal relaxation of the lattice are

Table 2. Summary of thermally stimulated exoemission peak values
in the synthetic, mineral, biological FAp series discussed
in the text.

SAMPLE	THERMALLY STIMULATED EXOEMISSION PEAK TEMPERATURES °C				
FA I			275	350	480-500 (combined OSEE/TSEE)
FA II		210	275	350	
FA I (preheated)		210	275	350	
DURANGO FAp	99	175		275	480
SHARK ENAMELOID	130		240-280 (double peak)		

NOTE : These temperatures are those of the stainless steel sample holder base not the sample surface

certainly associated with movement along the C-axis channels in
apatite, optical bleaching indicates that these defects cannot be
identical to those responsible for OSEE and OA although OS/TSEE would
suggest that the two are related.

These observations would extend the conclusions of Lapraz et al[31]
who stated that, in hydroxyapatites, traps responsible for TL appeared specific to the crystal lattice and independent of the chemical
impurities present.

It is noteworthy that while repeated irradiation and thermal
scanning improves the TSEE glow curve of human enamel, the TL glow
is reduced[54] probably due to adsorption of the emitted light caused
by a white/grey colour change in the sample.

Although no thermally stimulated relaxation technique is
suitable to identify the microscopic structure and chemical nature
of the emission centres involved[12], extension of the present work
into ultra-high vacuum (UHV) conditions will provide valuable
information concerning the surface reactivity of apatites. UHV
would allow differential measurement of charged particle species,
electrons, positive and negative ions during an on-going thermal
scan and provide dynamic surface information which could be compared
with the changing surface chemical composition using ESCA and other
analysis techniques. Exoemission can therefore be used to yield
important information concerning the electron and charged particle
availability on the surfaces of biological apatites. Release of
such charge carriers during defect annihilation will be of fundamental importance to the biological reactivity of this material.

ACKNOWLEDGEMENTS

    The work described above has only been possible with the
generous and helpful cooperation of a number of colleagues. I
should, in particular, like to express my thanks to Drs M. Petel
(Commissariat a l'Energie Atomique, CEN/FAR, France), G. Holzapfel
(Physikalisch-Technische Bundesanstalt, Berlin, FRG) (TSEE/OSEE),
R. Wild (C.E.G.B., Berkeley Nuclear Laboratories, England) (ESCA),
S.W.S. McKeever, K. Ahmed and P.D. Townsend (School of Mathematical
and Physical Sciences, University of Sussex, England) (3D-TL) and
J. Hay (DSC), Mrs J. Alves (TL) and Mrs S. Dipple (SEM) (Depts of
Chemistry, Physics and Materials Science respectively, University of
Birmingham, England) for the use of their apparatus. Financial
support from The Royal Society, London, the West Midlands Regional
Health Authority (for the new CPU-controlled apparatus), the supply
of samples from Professor J. Arends (Laboratory for Materia
Technica, University of Groningen), Drs J.C. Elliott and J.C.
Clement (the London Hospital Medical College, London) and preparat-
ion of the typescript by Gill Taylor are gratefully acknowledged.

REFERENCES

1.   "Physico-chemie et Crystallographie des apatites d'interet
      biologique". Colloques Internationaux CNRS No 230 Paris
      France (1975).
2.   D. Langdon, E. Dykes and R.W. Fearnhead, Defects, diffusion
      and dissolution in biological and synthetic apatite. Pages
      381-388 (1973) (Ref. 1).
3.   P.D. Johnson, Dichroic color centres in calcium fluorophosphate,
      J. Appl. Phys. 32:127 (1961).
4.   E.F. Apple, Observations on the formation of color centres in
      calcium halophosphates, J. Electrochem. Soc. 110:374 (1963).
5.   R.K. Swank, Colour centres in X-irradiated halophosphate
      crystals, Phys. Rev. 135(1A):A266 (1964).
6.   J.E. Davies, Preliminary results of the exoelectron properties
      of apatites, 5th Int. Symp. on Exoelectron Emission and
      Dosimetry, Zvikov, Czech.:203 (1976).
7.   P.D. Johnson, Some optical properties of powder and crystal
      halophosphate phosphors, J. Electrochem. Soc. 108:159 (1960).
8.   A. Boyde, Cutting teeth in the SEM, Scanning 1:157 (1978).
9.   D. Spitzer and J.J. Ten Bosch, The absorption and scattering
      light in bovine and human dental enamel, Calc. Tiss. Res. 17:
      129 (1975).
10.   J. Behari, S.K. Guha and P.N. Agarwal, Absorption spectra in
      bone, Calcif. Tiss. Res. 23:113 (1977)
11.   R.O. Becker and F.M. Brown, Photoelectric effects in human
      bone, Nature (Lond.) 206;1325 (1965).

12. P. Braunlich (Ed), "Thermally Stimulated Relaxation in Solids", Topics in Applied Physics Vol 37, Springer-Verlag, Berlin (1979).
13. R. Chen and Y. Kirsh, "Analysis of Thermally Stimulated Processes", Pergamon Press, Oxford (1981).
14. C.J. Maletskos, Thermoluminescence of bone, MIT Annual Report No. MIT 652-1:95 (1965).
15. J. Kastner, R. Hukkoo and B.G. Oltman, Thermoluminescence in bone, Argonne Nat. Lab; Rad. Phys. Div. Annual Report: 30 (1965).
16. F.L. Bordell and J. Kastner, Thermoluminescence in bone and bone-like materials, Argonne Nat. Lab; Rad. Phys. Div. Annual Report No. ANL-7489, Biology and Medicine, July 1967-June 1968.
17. D.G. Willhoit and A.D. Poland, Thermoluminescent characteristics of irradiated enamel and dentine, Health Phys. 15:91 (1968).
18. M. Jasinska and T. Niewiadomski, Thermoluminescence of biological materials, Nature (Lond.) 227:1159 (1970).
19. C. Christodoulides, Problems in TL dating bone, Proc. 8th Symp. Archeometry and Archeological Prospecting (1971)
20. C. Christodoulides, Thermoluminescent dating of materials of extraterrestial or biological origin, Ph.D. Thesis, University of Birmingham (1972).
21. C. Christodoulides and J.H. Fremlin, TL of biological materials, Nature 232:257 (1971).
22. H.S.T. Driver, The preparation of thin slices of bone and shell for thermoluminescence, Pact3:290 (1979).
23. L. Benko and L. Koszorus, Thermoluminescence dating of dental enamel, Nuc. Inst. and Meth. 175:227 (1980).
24. A.G. Wintle, Anomalous fading of thermoluminescence in mineral samples, Nature 245:143 (1973).
25. I.K. Bailiff, Use of phototransfer for the anomalous fading of thermoluminescence, Nature 264:531 (1976).
26. K.S.V. Nambi, Influence of rare earth impurities on TL characteristics, Pact 3:298 (1979).
27. J.E. Vaz, Effects of natural radioactivity on the thermoluminescence of apatite crystals at Cerro del Mercado, Mexico, Modern Geology 7;171 (1980).
28. J. Alves, J.E. Davies and S.A. Durrani, Thermoluminescence of fluorapatites and other mineral apatites: high-temperature emission and the effects of radiation damage, Proc. 3rd Int. Symp. on TL and ESR dating Elsinore, Denmark 26-31 July 1982 to be published in Pact 3: July 1983.
29. L. Suchow, Studies of color centres produced in apatite halophosphates by shortwave ultraviolet radiation. J. Electrochem. Soc. 108:847 (1961).
30. V.V. Ratnam, R. Jayaprakash and N.P. Daw, Thermoluminescence and thermoluminescence spectra of synthetic fluorapatite. J. Luminescence 21:417 (1980).

31. D. Lapraz, A. Baumer and P. Iacconi, On the thermoluminescence
     properties of hydroxyapatie Ca$_5$(PO$_4$)$_3$OH, Phys. Stat. Sol. (a)
     54:605 (1979).
32. S. Kolberg, S. Prydz and S. Dahm, Thermally stimulated
     luminescence in dental hard tissue and bone, Calc. Tiss. Res.
     17:9 (1974).
33. M.R. Chapman, A.G. Miller and T.G. Stoebe, Thermoluminescence
     in hydroxyapatite, Med. Phys. 6 (6):494 (1979).
34. H.S.T. Driver, personal communication.
35. Ist Int. Symp. on Exoelectron Emission, Innsbruc, Austria,
     Acta. Phys. Aust.10:313 (1975).
36. J. Kramer, "Der Metallische Zustand", Vanderhoek and Ruprecht,
     Gottingen (1950).
37. J. Kramer, Anwendung der Exoelectronen, Acta. Phys. Aust. 10:
     392 (1957).
38. H. Glafeke, "Exoemission" Chapter 5 in Ref (12).
39. G. Holzapfel and R. Nink, Zum ausseren Photoeffect on Electro-
     nen-haftzentren in kristallinen Festkorpern (Optisch stimul-
     ierte Exoelektronenemission - OSEE) P.T.B. - Mitt. 83:207
     (1977).
40. A. Bohun, The physics of exoelectron emission of ionic crystals,
     Proc. 3rd Int. Symp. on Exoelectrons, Braunschweig 3-12 July
     (1970).
41. G. Holzapfel, The evolution of volume concepts to describe
     exoelectron emission, 5th Int. Symp. Exoelectron Emission and
     Dosimetry Zvikov, Czech :19 (1976).
42. A. Scharmann and W. Kriegseis, Influence of surface parameters
     on exoelectron emission, 5th Int. Symp. Exoelectron Emission
     and Dosimetry Zvikov, Czech: 5 (1976)
43. L.I. Samuelsson, Mechanism for exoelectron emission mainly
     from LiF, Acta Radiobgica suppl. 359:1 (1979).
44. I.V. Krylova, Exoemission and the physico-chemistry of the
     surface - recent development, Proc. VIth Int. Symp. Exo-
     electron Emission and Applications, Ahrenshoop G.D.R.: 80 (1979)
45. L.G. Petersson, G. Frostell and A. Lodding, Secondary ion
     microanalysis of fluorine in apatites of biological interest.
     Z. Naturforsch 29:417 (1974).
46. G. Holzapfel, Zur exoelektronen-emission (Kramer-effekt) von
     berylliumoxid, Thesis, Technische Universitat Berlin (1968).
47. M. Petel, Recherches sur la dosimetrie par emissions
     electroniques stimulees, Thesis, Univ. Paul. Sabatier,
     Toulouse (No 318) (1976).
48. J.E. Davies and P. Ramsay, A microcomputer-controlled apparatus
     for simultaneous measurement of exoelectron emission and
     thermoluminescence, Proc. VIIth Int. Symp. on Exoelectron
     Emission and Applications Strasbourg , France, March 1983, to
     be published Rad. Proc. Dos. 4: (1983).

49. R. Nink and G. Holzapfel, Selective electron emission from f-centres in CsCl. J. de Phys. 3:19 (1973)

50. I.V. Krylova, The physico-chemical nature of exoelectron emission. 4th Int. Symp. Exoelectron Emission and Dosimetry Liblice Czech :145 (1973).

51. D.M. Todor, "Thermal Analysis of Minerals", Abacus Press (1976).

52. D.W. Holcomb and R.A. Young, Thermal decomposition of human tooth enamel, Calc. Tiss. Int. 31:189 (1980).

53. F.M. Ryan, R.C. Ohlmann, J. Murphy, R. Mazelsky, G.R. Wagner and R.W. Warren, Optical properties of divalent manganese in calcium fluorophosphate, Phys. Rev. B2(7):2341 (1970).

54. J. Alves, Ph.D. thesis, in preparation.

55. J.E. Davies, Exoelectron spectra of apatites of synthetic, mineral and biological orgin, Proc. VIth Int. Symp. Exoelectron Emission and Applications, Ahrenshoop G.D.R. (1979).

SURFACE CHEMISTRY OF SINTERED HYDROXYAPATITE: ON POSSIBLE RELATIONS WITH BIODEGRADATION AND SLOW CRACK PROPAGATION

(A Review)

K. de Groot
Dept. of Biomaterials
Schools of Dentistry and Medicine
Free University
Amsterdam
The Netherlands

1. INTRODUCTION

Ceramics with compositions resembling those of inorganic phases in bone and tooth tissue have been considered for clinical use since the end of the last century. Dreesman (1) applied Plaster of Paris ($CaSO_4$) to fill bony defects in patients. Since then, calciumphosphate has been advocated regularly as a suitable bioceramic to replace bone. However, it degrades too fast for new bone to fill the empty spaces left behind, so that it never became a widely accepted substitute for autologous bone. (2).

Somewhat later, but also a long time ago, calciumphosphate powders were used clinically, instead of fast resorbing calciumsulphate: Albee (3) reported the first clinical use of such a powder as early as 1920.

However half a century passed before calciumphosphate powders could be turned into blocks, usually by processes based on socalled 'sintering' (4). Now, early 1983, a number of producers are marketing several calciumphosphate products as a bone replacement and bone reconstructing material.

Although the biocompatibility of the two most used materials, sintered hydroxyapatite ($Ca_{10}(PO_4)_6(OH)_2$) and sintered β-whitlockite ($Ca_3(PO_4)_2$), is generally reported to be superior to all other bone replacement biomaterials, a number of properties need improvement. (5).

In this presentation we will discuss two of these properties, namely biodegradation and slow crack propagation (under tensile loading). We will pay special attention to the relation of these

97

properties with the surface chemistry of calciumphosphate salts in
general and of our sintered biomaterials of these salts in
particular.
The sequence of this presentation is as follows. Firstly, a short
review is given of calciumphosphate surface chemistry as pertinent
to our investigation secondly, biodegradation of and slow
crackpropagation in calciumphosphate bioceramics will be
summarized, and thirdly we will present some data that suggest the
importance of surface chemistry in understanding biodegradation
and slow crack propagation.

2. Surface chemistry of calciumphosphate salts used for sintering.

According to many authors, Driessens (6) being an example,
the probable phase composition of bone mineral is about 15% magnesium whitlockite (with a suggested formula $Ca_9Mg(HPO_4)(PO_4)_6$),
25% sodium and carbonate containing apatite (possibly
$Ca_{8.5}Na_{1.5}[(PO_4)_{4.5}(CO_3)_{1.5}]CO_3$ and 60% carbonated
octocalciumphosphate $(Ca_8(PO_4)_4(CO_3)(OH)_2)$.
Whether or not these percentage and formula's are correct, it is
obvious that the simplified view of bone having an inorganic phase
of pure apatite $(Ca_{10}(PO_4)_6(OH)_2)$ is not justified.
Since physiological processes are non-equilibrium processes, we
cannot expect equilibrium studies to represent the real situation.
Therefore, non-equilibrium phases, or transient phases, may be
present as well in the mineral phase. Especially mineral surfaces
will be composed of calciumphosphate salts both transient and in
thermodynamical equilibrium with surrounding physiological fluids.
Examples of thermodynamically stable salts are: brushite ($CaHPO_4$,
stable in aqueous solution of room temperature $-25^{\circ}C$ – at pH<5),
apatite ($Ca_{10}(PO_4)_6(OH)_2$, stable in aqeous solution of $25^{\circ}C$ at
pH>4,5), while an example of a salt not thermodynamically stable
at similar conditions is β-whitlockite ($Ca_3(PO_4)_2$, not stable at
roomtemperature in aqeous solution at any pH). This means that,
when one has obtained a powder by for example precipation
techniques with the total formula $Ca_3(PO_4)_2$, the equilibrium
composition cannot be that of β-whitlockite, but more probably a
mixture of brushite and apatite. Only heating to $800^{\circ}C$ would
change this mixture into a true β-whitlockite crystalline phase.
Cooling down rapidly will then give a thermodynamically unstable
β-whitlockite powder. This example shows that depending on heat-
treatment, one may have different endproducts after sintering,
even though the stoichiometry is the same.

3. Biodegradation of and slow crackpropagation in calciumphosphate bioceramics.

3.1. Biodegradation

In an earlier review (7) we have discussed biodegradation properties of calciumphosphate bioceramics, and hence a summary of the conclusion suffices here.
Like ordinary bone, calciumphosphate ceramics, implanted in bone will be subjected to remodelling, which means being degraded and replaced by new bone. Recent studies have elucidated that several steps can be identified in the biodegradation process: individual powder particles are loosened from the implanted ceramic which in turn can be ingested and intracellularly dissolved (Fig. 1), and a pure physicochemical, ionic dissolution can be envisaged.

The first step allows the following conclusions:

- since the rate of loosening of individual powder particles from the ceramic depends on the 'available' surface, increasing of pores large enough to allow cells into it (socalled 'macro-pores', pores with a diameter of at least 100 μ) will increase biodegradation rate.

- since the rate of loosening also depends on the 'neck' (i.e. the area of with which a particle is connected to another one), increasing 'microporosity', spaces left between sintered parti-cles, will also incrase biodegradation rate.

- the solubility rate of the material of which the 'neck' is made, will influence loosening rate of particles, and hence biodegradation, as well.

The latter factor, solubility rate, depends very much on surface chemistry of the bioceramic. For example, lattice structure, lat-tice vacancies, impurities such as Mg^{++}, Zn^{++}, F^-, protein absorb-tion in vivo, are variables influencing solubility rate of the surface and thus biodegadation.
The second step, pure physicochemical, ionic dissolution, depends similarly on factors, such as lattice structure, vacancies, impurities and protein adsorbtion.
As general conclusion we may state that biodegradation, both via particles loosened and subsequently digested cellularly and via direct ionic dissolution, is a phenomenon highly determined by surface properties of the implant concerned.

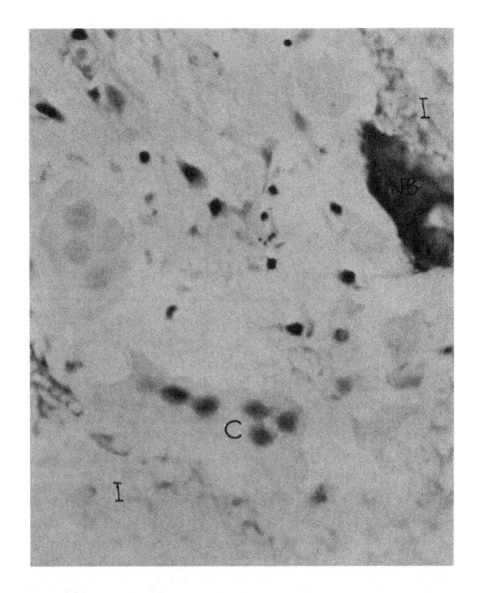

Fig. 1. Histological evidence of new bone (NB) being deposited against the walls of a large pore (diameter 100μ) in an implant (I). Cells (C) digest implantparticles from the pore surface.

3.2. Slow crackpropagation

Dense calciumphosphate ceramics (less than 5% porosity) have a compressive strength of about 600 ± 200 MN/m^2, while their tensile strength can be more than 150 ± 20 MN/m^2.
This seems very large, however, long term animal studies revealed that even with much lower tensile loads then 150 MN/m^2 fracture eventually occurred. (Fig. 2).
The phenomenon, that long term loading with sub-critical tensile loads may lead to fracture, is called static fatigue failure and encountered more or less in all ceramic materials.
Theoretically it is believed that the surface of ceramic contains small cracks, that grow slowly when continuously subjected to tensil loading. Interaction with humid environments accelerates the growth rate of these small cracks.
Without recalling the whole theory, the following formula (8) can be used to describe 'small crack growth' adequately:

$$\text{og } \log 1/P_s = C + \frac{m}{n-2} \log t_s + m \log \sigma_t \qquad 1$$

where: log = natural logarithm

m   = constant, socalled Weibull factor

n   = constant, determining susceptibility to fatigue failure

P_s = fraction of tested samples that survives a tensile load σ_t applied during a time t.

The equation shows the following:

-   if one keeps t_s constant (in a testing machine), the plotting log log $1/P_s$ versus log σ_t yields a straight line with slope m (= Weibull factor).

-   if one keeps σ_t constant (in animal studies for example), a straight line with slope $\frac{m}{n-2}$ is obtained when plotting log log $1/P_s$ versus log t_s.

Our experiments show in to be about 10-12, and $\frac{m}{n-2}$ about 1,0, so that n = 12.
Ceramics usually have Weibull (m) constants between 5 and 15, and n values of 10-100.
Other investigators found that for wet and dry calciumphosphate ceramics n equals about 10 and 30-50 respectively, so that our in-vivo experiments clearly indicate that slow crack propagation is the cause of failure.
Since values of n = 10 indicate serious fatigue failure problems, and n = 100 correlates with excellent resistance towards fatigue failure, we conclude not only that calciumphosphate ceramics are

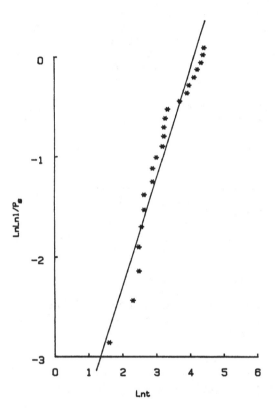

Fig. 2. Fracture versus natural logarithm of time (t in weeks) according to formula log log $1/Ps = C + \frac{m}{n-2} \log t + m \log \sigma_t$, σ_t assumed to be constant.

susceptible to fatigue failure, but also that interaction with
humid environments increases this problem. In this respect
calciumphosphate ceramics behave similar to oxide ceramics and
glasses.
The latter conclusion emphasizes the importance of surface properties in relation to fatigue failure of tensile loaded implants.

4. Preliminary research (9) on protein adsorption to surfaces of calciumphosphate ceramics

Since implants are used in a non-equilibrium environment,
made of a complex system of physiologic components, we have
carried out protein adsorbtion studies. To do so, we incubated
various calciumphosphate components in serum, and studied protein
adsorbtion both quantitatively and qualitatively. As Table I
shows, we found that amount and mutual ratios of adsorbed proteins
were different for all the calciumphosphate ceramics tested. Not
only crystallographic structure (β-whitlockite versus apatite) but
also sinter procedure (heat-treatment) influenced protein adsorbtion and, hence, surface properties. This shows, that the various
calciumphosphate surfaces are not in thermodynamical equilibrium
with the surrounding physiological fluids, but are covered with
proteinlayers under conditions described by non-equilibrium
thermodynamics (as are most biological processes).
The conclusion is therefore that non-equilibrium thermodynamics
are the key to understanding calciumphosphate surface chemistry in
physiological conditions. This may seem trivial, but many
investigators (6) study surface compositions under in vivo
equilibrium conditions, and assume implicitly the thus obtained
information to be true for dynamic biological processes as well.
In terms of this presentation it means that biodegradation and
fatigue failure in vivo should be studied not only under
physiological conditions but also with the dynamic character of
relevant adsorbtion phenomena in mind.

Table I. Example of protein adsorption differing in absolute amount and ratio (for two proteins).

Powders		% Adsorption		
		C3 Mean ± SD (n=5)		α_2-HS Mean ± SD (n=5)
$CaHPO_4$	(Brushite)	61	7	39	3
$Ca_{10}(PO_4)_6(OH)_2$	(HA)	59	9	53	6
$Ca_3(PO_4)_2$	(TCP)	62	6	56	10
Sintered HA	800°C	72	4	nd	
Sintered HA	1150°C	45	7	nd	
Sintered TCP	1300°C	3	5	nd	
Al_2O_3	1150°C	39	4	35	11

(nd = not detectable).

CONCLUSION

Biodegradation of and slow crack propagation in bioceramics made of calciumphosphate salts are phenomena strongly influenced by the surface chemistry of these implant materials. Since we have shown that adsorbtion of proteins and presumably other (ionic) compounds cannot be described by equilibrium thermodynamics, we believe that non-equilibrium thermodynamics, representing better the nature of biological phenomena, should give better understanding of differences in biodegradation and slow crack propagation.

REFERENCES

1. H.Dreesman, Uber Knochenplombierung, Beitr.Klin.Chir. 9: 804 1894.
2. J.W.Frame, Resorption characteristics of calciumsulphate implants, J.Dent.Res. 53: Abstr. 81, 1974.
3. F.H.Albee, Studies in Bonegrowth, Ann.Surg. 71: 32 1920.
4. R.Pampuch, Ceramic Materials, Elsevier Amsterdam, 1976.
5. K.de Groot, Presentation NIH Consensus Conference on Clinical Biomaterials, Bethesda, 1982 (Nov.1-3).
6. F.C.M.Driessens, Probable Phase Composition of the mineral in bone, Z.Naturforsch 35c: 357 1980.
7. K.de Groot, Degradable Ceramics, in: Biocompatibility of Clinical Implant Materials, Vol I, D.F.Williams, Ed., CRC Press, Boca Raton, USA, 1981.
8. B.J.Dalgleish, R.D.Rawlings, A Comparison of the mechanical behavior of aluminas in air and simulated body environments, J.Biomed.Mater.Res. 15: 527 (1981).

ADSORPTION OF N,N-DIMETHYL-P-AMINOPHENYLACETIC ACID
ON HYDROXYAPATITE

Dwarika N. Misra

American Dental Association Health Foundation
Research Unit at the National Bureau of Standards
Washington, D.C. 20234

ABSTRACT

The adsorptive properties of N,N-dimethyl-p-aminophenyl-acetic acid on hydroxyapatite have been investigated. It is a fast-acting amine polymerization accelerator, but tensile strengths of composites of resin filled with apatite show that it is not an effective coupling agent for a hydroxyapatite-dental resin composite.

INTRODUCTION

N,N-Dimethyl-p-aminophenylacetic acid was recently found to be a fast-acting amine accelerator for the ambient polymerization of dental composites[1-4]. In low concentrations this amine imparts good color stability to polymers[1]. The amine would be expected to be surface-active with apatitic surfaces because it contains a carboxylate group[5-7]. Conceivably, it might act as a coupling agent for tooth structure if it bonded covalently to the polymer. However, it is likely that the bonding between the amine and apatitic mineral may not be hydrolytically stable because the compound neither possesses an effective hydrophobic moiety nor multiple surface-bonding groups which would add stability to the bonding.

To test the above hypotheses about the new amine its adsorption onto hydroxyapatite was studied from solutions prepared in ethanol (95%) or methylene chloride. Like benzoic[6] and anthranilic[5] acids, the new amine, which is also an acid, is adsorbed reversibly from ethanol (95%) and irreversibly from methylene chloride solutions. The irreversible adsorbate, however, is not hydrolytically stable since it is completely washed off by excess water (or ethanol). At maximum adsorption, the configuration of the reversibly adsorbed molecules is flat with respect to the benzene ring and that of the irreversibly adsorbed molecules is upright. Matching of the experimental areas with the effective areas based on molecular models suggests that the adsorbate molecules in both configurations are anchored to the surface by carboxylate groups, about which they may be slowly rotating.

The tensile strengths of polymerized dental resin containing N,N-bis(2-hydroxyethyl)-p-toluidine (DHPT) or the new amine in the monomer paste with hydroxyapatite as a filler are comparable. It may, thus, indicate that the new amine does not act as an effective coupling agent. The strengths are also not affected whether the composite is kept dry or wet.

MATERIALS AND METHODS*

Hydroxyapatite

The apatite was tribasic calcium phosphate [Fisher certified, C127, with a chemical formula given as approximately $Ca_{10}(OH)_2(PO_4)_6$]. It was repeatedly washed with boiling water before use; the physical and chemical details of its preparation have been described elsewhere[8]. It had a surface area (BET, N_2) of 41 m^2/g. The amount of physically adsorbed water (1.57%, about 1.5 monolayers) on the apatite was determined by evacuating (at 100 N/m^2) the weighed samples at 105°C for several hours and then weighing after dry air was introduced into the vessel.

N,N-dimethyl-p-aminophenylacetic acid

The synthesis of the new amine has been described[1]. It was recrystallized from methylene chloride solution (mp = 112°C). The NMR and IR analyses confirmed the structure.

U.S.P. grade 95% ethanol (190 proof, containing 5% water) was obtained from Publicker Industries, Inc., Philadelphia, PA. The methylene chloride was Fisher certified reagent grade. BIS-GMA (Nupol) was obtained from Freeman Chemical Corporation,

---

*Certain commercial materials and equipment are identified in this paper to specify the experimental procedure. In no instance does such identification imply recommendation or endorsement by the National Bureau of Standards or the American Dental Association Health Foundation, or that the materials and equipment identified are necessarily the best available for the purpose.

Port Washington WI; triethyleneglycol dimethacrylate from Sartomer Co., West Chester, PA; butylated hydroxytoluene from Eastman Chemical Products, Inc., Kingsport, TN; benzoyl peroxide (Lucidol 98) from Pennwalt Corp. Buffalo, NY; and N,N-bis(2-hydroxyethyl)-p-toluidine (DHPT, recrystallized from toluene-ligroin) from Mobay Chemical Co., Pittsburgh, PA.

Reversible adsorption and desorption

The hydroxyapatite samples (1.000 g each) were shaken with a series of standard solutions (10 mL each) of the new amine in ethanol (95%) at room temperature (23.0 ± 0.5°C) for approximately 30 minutes, a period observed to be sufficient for attainment of equilibrium. The slurry was filtered through a medium-pore fritted disc by momentarily applying a light suction. The concentration of the adsorbate in the filtrate was determined from its absorbance (at 325 nm) on a double-beam spectrophotometer (Coleman model 124 D) and a standard Beer's law plot. Spacers were used for concentrated solutions. The amount adsorbed, A (mol/g), is derived by the relation: $A = V \Delta C/W$, where V(L) is the volume of solution in contact with W g of the adsorbent and ΔC (mol/L) is the difference of the initial and final concentrations of solution. The adsorption values were reproducible within ± 2.5%. The control experiments showed that the amounts adsorbed did not change with time (up to 7 days) after 30 minutes.

The adsorbate could be completely removed by repeatedly washing the apatite with an excess of pure solvent (100 - 150 mL) after the equilibrium was reached. The desorbed amount was calculated by measuring the total volume of wash solution and its concentration spectrophotometrically.

Irreversible adsorption and desorption

The adsorption of the new amine from methylene chloride was similarly determined (absorbance measured at 330 nm). The adsorption was total and exhaustive (i.e. all adsorbate was removed) from dilute solutions and constant from concentrated solutions.

The adsorbate could not be washed off with an excess of neat methylene chloride. The desorption was considered complete when final washings showed no absorbance. When a dried sample of the apatite containing a known amount of the new amine was washed with ethanol (95%) or water, all adsorbate was removed from the surface, as determined from the volume and concentration of the filtrate.

Composite preparation and strength

The specimens were prepared by mixing two pastes (accelerator or initiator) consisting of the hydroxyapatite and methacrylate monomers which contained either amine or peroxide. The amine coated apatite powder was directly mixed with the peroxide paste. The monomers were BIS-GMA (69.95% by weight) and triethyleneglycol dimethacrylate (29.95%), with butylated hydroxytoluene (0.10%) as added stabilizer*. The polymerization accelerators were DHPT (0.54%, based on stabilized monomer) or the new amine (0.49%, 0.25%) and the peroxide was benzoyl peroxide (1% in the initiator paste). In addition, a powder/paste formulation technique was used to prepare specimens from hydroxyapatite containing the adsorbed new amine (see Table 1). In all cases, the monomer to filler ratio was 1.2 by weight.

Table 1. Diametral Tensile Strengths of Polymer Filled with Hydroxyapatite

Composite[a]	Amine Used[b] (Conc.,% by Wt)	Setting Time[d] Min	No. of Samples	Mean Tensile Strength (Std. Deviation), MPa
OHAP	DHPT (0.54)	4.5	Dry, 5	27.2 (2.6)
			Wet, 5	25.5 (2.7)
OHAP	New amine (0.49)	1	Dry, 5	26.9 (1.4)
			Wet, 6	26.5 (1.9)
OHAP	New amine (0.25)	2	Dry, 5	24.5 (2.5)
			Wet, 6	24.1 (3.0)
OHAP-coated[c] (0.45)	New amine (0.375)	3	Dry, 5	25.0 (1.8)
			Wet, 6	25.0 (2.2)
OHAP-coated[c] (0.18)	New amine (0.15)	5	Dry, 5	19.6 (2.6)
			Wet, 5	19.9 (2.1)

[a]Monomer to hydroxyapatite (OHAP) ratio was 1.2 by weight.

[b]Amine in solution in its monomer paste. In OHAP-coated samples, the new amine was adsorbed on the apatite and no amine monomer paste was used, and the concentration of amine in monomer was calculated on the basis that all amine desorbed into monomer.

[c]The amount in parenthesis is % of amine adsorbed on apatite by weight (one monolayer amount is 1.92%).

[d]According to ADA Specification No. 8.

[e]Wet samples were stored in water and dried in air in closed bottles at 37°C for 28 days before testing.

---

*The monomers may also have contained unspecified stabilizers as received.

    The specimens for diametral tensile strength measurements
were molded in a stainless steel die (diameter = 6.1 mm, thickness = 3.2 mm) as described[9] in The American Dental Association
Specification No. 9. They were stored in water or air at 37°C
for 28 days before their strengths were measured on an Instron
testing instrument at a cross-head speed of 0.5 cm/min.

RESULTS

    The adsorption isotherm of the new amine on hydroxyapatite
from ethanol (95%) at 23°C is shown in Figure 1. The
reversibility of adsorption is confirmed by a complete removal of
the adsorbate with an excess of ethanol (95%) (Figure 1).

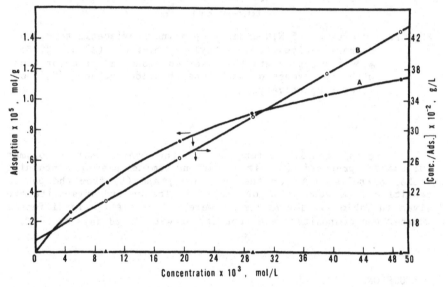

Fig. 1.   Adsorption of N,N-dimethyl-p-aminophenylacetic acid on
          hydroxyapatite from ethanol (95%) at 23°C: (A) isotherm
          (●), (▲ the adsorbed amount after desorption with an
          excess of pure ethanol; (B) the Langmuir plot (o), the
          straight line is obtained by linear regression.

    The adsorption isotherm of the new amine from methylene
chloride (Figure 2) is typical of an irreversibly adsorbed
solute; a total adsorption (i.e. removal of all adsorbate) from
dilute solutions and a constant adsorption (10.7×10^{-5} mol/g)
from concentrated ones. The adsorbate is not desorbed by washing
with an excess of methylene chloride but it could be completely
removed by washing with ethanol (95%) or water (Figure 2).

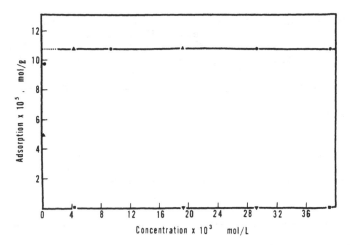

Fig. 2. Adsorption of N,N-dimethyl-p-aminophenylacetic acid on
hydroxyapatite from methylene chloride (●) at 23°C:
▲,▼or ■ represents the adsorbed amount after desorption
with an excess of methylene chloride, ethanol (95%) or
water respectively.

Diametral tensile strengths for the hydroxyapatite-polymer
composites prepared (a) from various amine pastes containing
either dissolved DHPT or the new amine, and (b) from the powder
resulting from adsorption of the new amine on hydroxyapatite are
given in Table 1. The strengths were not statistically different
whether the composites were kept dry or wet for 28 days at 37°C.

DISCUSSION

Like anthranilic[5] or benzoic[6] acids, the new amine is
reversibly adsorbed from ethanol (95%) and irreversibly from
methylene chloride solutions onto hydroxyapatite. This behavior
is to be expected since the new amine is also a carboxylic acid.
The solvent-dependent reversibility or irreversibility of an
adsorbate is perhaps linked to whether the solvent can develop
strong hydrogen bonding with the adsorbate and/or the substrate.
This does not mean to imply that the forces holding the adsorbate
molecules onto the surface are exclusively hydrogen bonds;
dipole-dipole and/or ionic forces may also be involved. The

study of adsorption isotherms cannot uniquely resolve this issue by itself. Other physical measurements (e.g. infrared spectroscopy of adsorbed species) must be pursued to elucidate the nature of the adsorptive forces.

The adsorbed N,N-dimethyl-p-aminophenylacetic acid, like anthranilic[5] and benzoic[6] acids, would not be expected to be hydrolytically stable since the compound neither possesses an effective hydrophobic moiety nor multiple surface-bonding groups. Three compounds that are irreversibly adsorbed onto hydroxyapatite and are not removed by water are NPG–GMA[10], an adduct of the diglycidyl ether of bisphenol A with N-phenylglycine[11], and zirconyl methacrylate[12]; probably because these compounds possess hydrophobic moieties. It would appear that, for NPG–GMA and the bisphenol A adduct, "surface chelation" may play a significant role and, in general, thermodynamic factors, e.g. entropic and enthalpic parameters, may also be important.

The reversible adsorption of the new amine from ethanol (95%) may be represented by the Langmuir equation[13]:

$$\frac{C}{m} = \frac{C}{M} + \frac{1}{bM} \qquad (1)$$

where C is the equilibrium concentration (mol/L), m the adsorbed amount (mol/g), M the saturation amount of adsorbate (mol/g), and b a constant related to the heat of adsorption (L/mol). The Langmuir plot is shown in Figure 1 and its linearity is very good. The reciprocal of the slope of the plot gives the value of the saturation amount of the adsorbate and the constant b is obtained by dividing the slope by the intercept (Table 2).

The effective cross-sectional area, σ, of the adsorbate molecule may be calculated from the saturation amount (M),

$$\sigma = S/NM \qquad (2)$$

where S is the surface area of adsorbent and N is Avogadro's number (Table 2, column 7). The molecular areas thus obtained may be related to their orientations and conformations on the adsorbent surface. These areas may also be obtained from Fisher–Hirschfelder–Taylor scale models. At room temperature, the adsorbed molecule, if it is not restricted, would tend to rotate slowly on the surface. The effective cross-sectional area of a rotating molecule (radius = r) is $2\sqrt{3}\ r^2$ (area of a hexagonally packed circular disc + interstitial area). These areas are given in Table 2 (footnote d).

Table 2. Adsorption isotherm constants and certain comparisons

| Adsorbate | Solvent[a] | Langmuir plot consts.[b] | | Amt. at saturation x 10^7, mol/g | Heat term (b), L/mol | Area/molecule (σ)[c,d], Å² |
		Slope x 10^{-4} g/mol	Intercept x 10^{-2} g/L			
New amine	Ethanol (95%)	5.65	15.48	1.77	36.51	385
	CH_2Cl_2	--	--	10.70	--	64
Benzoic acid	Ethanol (95%)	2.62	22.10	3.82	11.86	178
	CH_2Cl_2	--	--	17.20	--	40
Anthranilic acid	Ethanol (95%)	0.88	10.70	11.40	8.20	60
	CH_2Cl_2	--	--	17.20	--	40

[a]The adsorption is reversible from ethanol (95%) and irreversible from CH_2Cl_2.

[b]The correlation coefficient in each case is greater than 0.99.

[c]σ = S/NM (Eq. 2).

[d]Based on molecular model, the effective area of the new amine molecule: lying flat and rotating about itself = $2\sqrt{3}(10.5)^2 = 382$ Å², and rotating about carboxylate group in an upright position = $2\sqrt{3}(4.3)^2 = 64$ Å². The corresponding pair of values for benzoic and anthranilic acids are: 175 Å², 40 Å²; and 60 Å², 40 Å² respectively.

Surface configuration of N,N-dimethyl-p-aminophenylacetic acid, like benzoic and anthranilic acids, adsorbed from the same solvents on hydroxyapatite, is identical (Table 2, footnote d). In addition to hydrogen bonding with the hydroxyapatite surface, the carboxylate group, even if not fully ionized, would have some ionic and dipolar interactions with the ionic substrate. Water molecules, unlike nonprotic solvents like methylene chloride, can substitute these bonds and/or interactions and desorb these adsorbates. This may be a simple explanation for the reversibility or irreversibility of these adsorbates from these solvents. The orientational behavior of the adsorbate molecules may be understood if it is considered that the interaction of the benzene ring in protic solvents may be stronger with the surface than with the solvent, whereas the reverse may be true in nonprotic solvents.

Since the tensile strengths of composites using DHPT or the new amine, whether wet or dry, are similar under the experimental conditions (Table 1), chemical and/or strong physical bonding does not appear to develop between the new amine and the polymer matrix. It makes no significant difference in strengths whether the new amine is used as dissolved in monomer paste or coated on hydroxyapatite. A perceptible increase in the composite tensile strengths is observed with an increase in concentration of amine in the system (Table 1).

CONCLUSIONS

N,N-Dimethyl-p-aminophenylacetic acid, like benzoic and anthranilic acids, is adsorbed reversibly on hydroxyapatite from ethanol (95%) and irreversibly from methylene chloride. The reversible isotherm follows the Langmuir plot.

The surface configuration of the reversibly adsorbed molecules is flat with respect to the benzene ring and that of the irreversibly adsorbed molecules is upright. The adsorbate molecules in both configurations are anchored to the surface by carboxylate groups, and they may be slowly rotating about them.

The new amine does not act as an effective coupling agent between the dental resins and hydroxyapatite composites since the tensile strengths of the composites are not increased as compared to the composites made with DHPT.

ACKNOWLEDGEMENT

This investigation was supported in part by Research Grant DE05129-05A1 to the American Dental Association Health Foundation from the National Institutes of Health-National Institute of Dental Research, and is part of the dental research program conducted by the National Bureau of Standards in cooperation with the American Dental Association Health Foundation.

REFERENCES

1. D. M. Dulik, Evaluation of Commercial and Newly-synthesized Amine Accelerators for Dental Composites, J. Dent. Res. 58:1308-1316, 1979.
2. G. M. Brauer, D. M. Dulik, J. M. Antonucci, D. J. Termini, and H. Argentar, New Amine Accelerators for Composite Restorative Resins, J. Dent. Res. 58:1994-2000 (1979).
3. J. M. Antonucci, D. N. Misra, and R. J. Peckoo, The Accelerative and Adhesive Bonding Capabilities of Surface-Active Accelerators, J. Dent. Res. 60:1332-1342 (1981).
4. G. M. Brauer, J. W. Stansbury, and J. M. Antonucci, 4-N,N-Dialkylaminophenethanols, -Alkanoic Acids and Esters: New Accelerators for Dental Composites, J. Dent. Res. 60:1343-1348 (1981).
5. D. N. Misra, and R. L. Bowen, Adsorptive Bonding to Hydroxyapatite I: Adsorption of Anthranilic Acid—the Effect of Solvents, Biomaterials 2:28-32 (1981).
6. D. N. Misra, Adsorptive Bonding to Hydroxyapatite II: Adsorption of Benzoic acid: Role of Solvent and Hydrogen Bonding, (in preparation).
7. M. Kresak, E. C. Moreno, R. T. Zahradnik, and D. I. Hay, Adsorption of Amino Acids onto Hydroxyapatite, J. Colloid Interface Sci. 59:283-292 (1977).
8. D. N. Misra, R. L. Bowen, and B. M. Wallace, Adhesive Bonding of Various Materials to Hard Tooth Tissues VIII: Nickel and Copper Ions on Hydroxyapatite; Role of Ion Exchange and Surface Nucleation, J. Colloid Interface Sci. 51:36-43 (1975).
9. Guide to Dental Materials and Devices, 6th ed, American Dental Association, 1972-1973, p. 194.
10. D. N. Misra, and R. L. Bowen, Adhesive Bonding of Various Materials to Hard Tooth Tissues XII. Adsorption of N-(2-Hydroxy-3-Methacryloxypropyl)-N-Phenylglycine (NPG-GMA) on Hydroxyapatite, J. Colloid Interface Sci. 61:14-20 (1977).
11. D. N. Misra, and R. L. Bowen, Adhesive Bonding of Various Materials to Hard Tooth Tissues XI: Chemisorption of an Adduct (of the Diglycidyl Ether of Bisphenol A with N-Phenylglycine) on Hydroxylapatite, J. Phys. Chem. 81:842-846 (1977).
12. D. N. Misra, Unpublished results.
13. A. W. Adamson, "Physical Chemistry of Surfaces," Interscience, New York, (1982) p. 388.

ADSORPTION OF PHOSPHONYLATED

POLYELECTROLYTES ON HYDROXYAPATITE

H. Ralph Rawls* and Israel Cabasso**

*Louisiana State University Medical Center
New Orleans, LA 70119

**The Polymer Research Inst. of State University of N.Y.
College of Environmental Science and Forestry
Syracuse, New York 13210

ABSTRACT

The adsorption of phosphonylated polyphenylene oxide onto hydroxyapatite was investigated as a basis for biomedical applications involving surface modification of teeth and bone. The half-acid, half-ester form of the polymer was found to adsorb rapidly and the rate and amount increased with concentration. Phosphate, but not Ca^{++}, is released into solution during adsorption. The ratio of released phosphate to polymer-bound phosphonate absorbed from solution, is about 2.5 at very low initial polymer concentrations but decreases rapidly to about 0.5 with increased concentration. For equivalent ionic strengths, added salts increase the rate and amount of polymer adsorbed in the order: $CaCl_2 > KCl = KNO_3 > K_2HPO_4 >$ no salt. Desorption is enhanced by K_2HPO_4. These results can be explained by an adsorption mechanism that is controlled by the degree of charge neutralization along the polymer chain. Intrachain charge repulsion results in an extended random-coil conformation and adsorption in a thin layer with binding at many polymer sites. Charge screening, neutralization and chelation cause the random-coil to collapse. A collapsed conformation occupies only a relatively few binding sites and thus adsorption results in the formation of a thick layer. The conformation and binding of adsorbed layers are expected to be of major importance in surface-controlled phenomena such as adhesion, drug delivery, and flocculation.

INTRODUCTION

Organic phosphates and phosphonates are well known to have high affinities for adsorption to mineral surfaces, including the mineral phases of teeth and bone.[1-3] These compounds interfere with crystallization and dissolution processes and inhibit dental calculus formation and caries.[1,4,5] Because organic phosphate esters, $PO(OR)_3$, are susceptible to complete hydrolysis, most attention has focused on the phosphonate derivatives, $RPO(OR)_2$, which are more stable and have a P-C bond that is not readily disrupted. Binding of the diphosphonates[6] has been found to be less readily reversible because the simultaneous detachment of both groups has a low probability. Thus, it can be expected that polysubstituted phosphonates will be essentially permanently bound.[7] A polymer chain having pendent phosphonate groups, as recently reported,[8] is such a compound. These polymers, designated "polyphosphonates," when adsorbed to hydroxyapatite alter surface and transport properties and may prove beneficial in a number of biomedical applications involving teeth and bone: for example, adhesion, protection against mineral loss, and site specific drug delivery.

Polymers adsorb from solution in a variety of patterns that range from attachment by a large number of groups along the polymer chain, to attachment by only a few with the rest of the chain extending into solution as loops and tail segments.[9] As a basis for eventual biomedical and other applications, we are studying the factors that affect polyphosphonate adsorption onto hydroxyapatite. In the present study we report on some of the effects of polyelectrolyte and ion concentration on the adsorption mechanism.

MATERIALS AND METHODS

Polymers. The polyphosphonate was prepared by substitution of poly(2,6-dimethyl-1,4-phenylene oxide), PPO ($M_w \approx 30,000$, General Electric grade 691-111). After bromination of the ring and methyl groups, the Abruzov reaction was utilized to form the phosphonate methyl ester, as described earlier.[10] The ester was then hydrolyzed in alkaline dimethyl formamide and precipitated in HCl to produce the half-ester, half-acid polymer.

The degree of polymerization was estimated from the known value, approximately 300, of the starting polymer.[10] The degree of substitution was determined by NMR to be 1.2 phosphonates per repeating unit, thus $X \approx 240$ and $Y \approx 60$. The half ester/half acid form of the polymer was designated PPOBr P-1.2A. Titration of 0.3mM PPOBr P-1.2A (concentration based on phosphonate groups) with 0.09mM KOH was used to determine the degree of hydrolysis of the phosphonate groups. The degree of ionization at any given pH was

determined from the titration curve. Infrared spectroscopy and
microanalysis were used to verify the structure shown below:

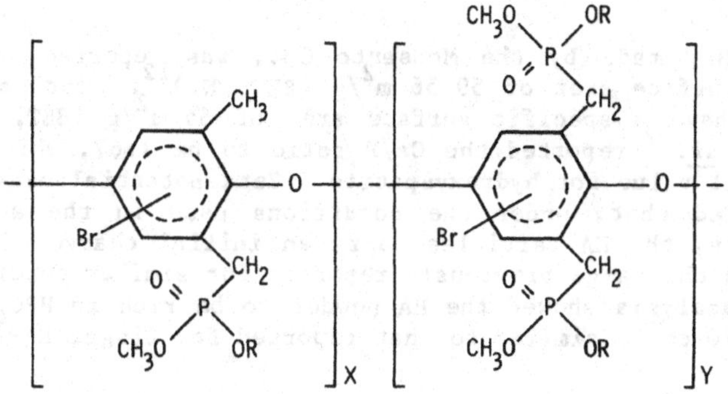

PHOSPHONYLATED POLYPHENYLENE OXIDE
PPOBr$_\phi$P (R = H, CH$_3$)

Polymer solution viscosity was measured at 10g/l by the
Ubbelohde capillary flow method at 25 C. The specific viscosity was
determined and divided by concentration to derive the reduced
viscosity.

Adsorption studies. The acid form of the polymer,
PPOBr$_\phi$P-1.2A, was dissolved in a small volume of KOH solution,
diluted with water, dialyzed against deionized water to remove all
salts, and then brought to pH 7 with a small amount of HCl. These
polymer solutions, of varying concentrations, were then combined
with hydroxyapatite (HA) to form a slurry density of 1g HA/l and
agitated at 100 RPM at 23-24C. At various intervals aliquots were
withdrawn and separated by either filtration (0.45 micron membrane)
or by centrifugation. Aliquots were analyzed spectrophotometrically for decrease in polymer concentration (phenyl-group
absorption at 283 nm), and for increases in phosphate[11] and calcium
(atomic adsorption). HA is very slightly soluble at pH 7. Therefore
phosphate determinations were also made on control slurries that
contained no polymers. These phosphate values were subtracted from
corresponding values observed during adsorption. Thus the
phosphate concentrations reported are considered to be due only to
adsorption effects.

In some solutions certain salts were included: KCl or KNO$_3$ (as non-specific, "indifferent," electrolytes), KHPO$_4$ (as a phosphate ion source), or CaCl$_2$ (a divalent ion). The salts were present in amounts at least equivalent in ionic strength to that of the polyelectrolyte. Turbidity measurements were used to determine the conditions under which precipitation occurs in the presence of Ca^{++}.

The HA, made by the Monsanto Co., was reported to have a specific surface area of 59.56 m^2/g (BET, N$_2$)[12]. Our sample was found to have a specific surface area of 65 m^2/g (BET, argon)[8]. Kresak et al.[12] reported the Ca/P ratio to be 1.67, which is the theoretical value for hydroxyapatite. Zeta potential measurements demonstrated that, under the conditions used in the adsorption experiments, the HA particles carry an initial charge of +16 mV[13]. This is in the range previously reported for similar conditions. Infrared analysis showed the HA powder to be rich in HPO$_4$ (strong band at 870 cm^{-1}) similar to that reported for Biogel HAP.[13]

RESULTS

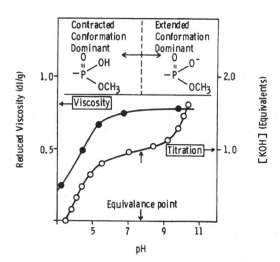

Figure 1. Effect of pH changes on the viscosity and degree of neutralization of a phosphonylated polyelectrolyte.

Polymer identification. After hydrolysis, the polyphosphonate shown above was determined to be 46% hydrolyzed. There are approximately 290 R-groups = H, and approximately 70 = CH_3 in the structure shown.

pH dependence. The solution viscosity of PPOBr$_\phi$P-1.2A increases rapidly with pH, until about pH 7, beyond which little further increase is observed (Fig. 1). Titration of this polymer demonstrated that it is a weak acid with pKa = 4.4 and a stoichiometric equivalence point at pH = 7.5 (Fig. 1). It is therefore of similar acidity as acetic acid (pKa = 4.75). At pH 7, 95% of the phosphonic acid groups are ionized. This accounts for the solution viscosity behavior: as pH 7.5 is approached the increasing number of charges along the chain repulse each other and progressively extend the random coil conformation, which maximizes viscosity. Conversely, below pH 7.5 charge repulsions diminish, the chains contract and occupy less volume, and a reduction in viscosity is observed. This is typical for polyelectrolytes.[14]

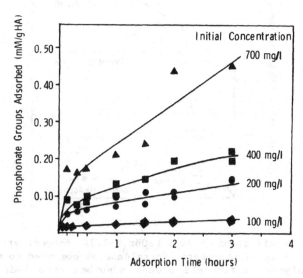

Figure 2. The effect of initial concentration on the rate of adsorption of PPOBr$_\phi$P-1.2A onto hydroxyapatite from water at pH 7.

Adsorption. The rate and amount of polymer adsorption depends on initial concentration, as shown in Fig. 2. During adsorption, HA releases phosphate into solution, but not Ca^{++}. At an initial polymer concentration of 100mg/1, phosphate is released into solution at a relatively slow rate, as shown in Fig. 3. At this concentration the 3 hr amount of phosphate released is more than twice the molar amount of phosphonate groups adsorbed. At higher initial polymer concentrations the release of phosphate is very rapid and is completed within a few minutes. Adsorption also increases and the ratio of phosphate to phosphonate reverses: twice as many phosphonate groups are adsorbed as phosphate ions are released. This is shown in Fig. 4 where the phosphate/phosphonate ratio (PO_4/PO_3) is plotted versus adsorption time for initial polymer concentrations ranging from 100 to 700 mg/1.

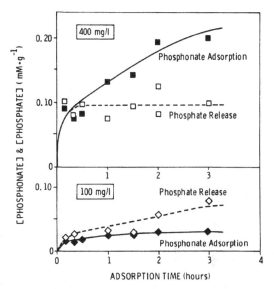

Figure 3. The effect of initial PPOBr,P-1.2A concentration at pH 7 on the rate of phosphate ion release as compared to the number of polymer phosphonate groups adsorbed onto hydroxyapatite from water.

In a separate experiment it was observed that the HA powder used in these experiments is slightly soluble and requires about 3 hr to reach equilibrium. This calls for the presence of Ca^{++} in the aqueous polymer solutions during adsorption. The fact that Ca^{++} was not detected indicates that it must be bound by the phosphonate groups.[8] We had previously observed that $CaCl_2$ does not cause rapid precipitation of the phosphonylated polymer unless the calcium concentration is in many-fold excess of the HA "saturation level ."

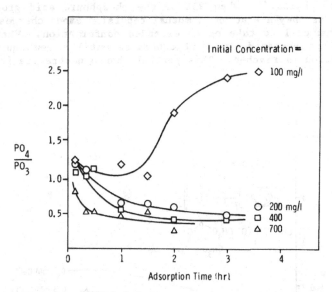

Figure 4. The effect of initial PPOBr,P-1.2A concentration at pH 7 on the ratio of phosphate-released to phosphonate-groups-adsorbed onto HA after 3 hrs.

Nor does Ca^{++} cause precipitation unless it is present in many-fold excess of the number of polymer-bound phosphonate groups.[8] In an experiment in which Ca^{++} (as $CaCl_2$) was added to polyphosphonate solutions at a level corresponding to HA saturation, 15 hr was required before turbidity was detectable. Thus, simple precipitation in the bulk of the solution does not seem to account for the failure to observe an equilibrium plateau at polymer concentrations above about 200 mg/l when the slurry density is 1 g/l (Fig. 2). Rather, the presence of phosphate, calcium and hydroxyl ions due to HA dissolution during adsorption must effect an enhancement of adsorption. This was tested by the addition of salts containing Ca^{++} and $HPO_4^=$ ions to polyelectrolyte solutions.

<u>Ion effects on adsorption.</u>   The results of adsorption experiments in the presence of various added ions, are shown in Fig. 5.   KCl and KNO_3 (not shown), as "indifferent" electrolyte controls, demonstrate that the electrolytic strength of the solution is an important parameter.   The addition of KCl or KNO_3 enhances adsorption compared to deionized water solutions. This is likely due to the screening effect of the electrolyte on the charges along the chains.[14]   When 95% of the phosphonic acid groups are ionized, as they are at pH 7, mutual repulsion among charges causes the random coil to take on an extended conformation. When these charges are screened the coil contracts until a new equilibrium conformation is reached. This partial charge neutralization would

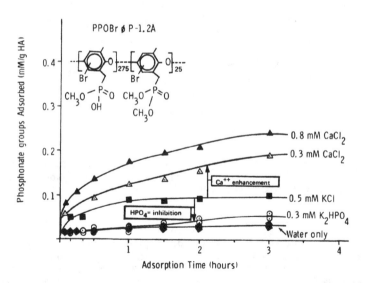

Figure 5. The effect of simple electrolytes on the adsorption pattern of PPOBr ϕ P-1.2A at 100 mg/l (0.3mM based on phosphonate groups) in the presence of HA at pH 7.

reduce <u>intercoil</u> repulsions and thereby permit a denser packing at the surface during adsorption.[8] Thus both adsorption rate and amount adsorbed would be expected to increase.

Calcium ions present during adsorption cause a marked increase in both adsorption rate and amount adsorbed, while phosphate has a smaller effect (Fig. 5). The order of adsorption enhancement, for equivalent ionic strengths, is: $CaCl_2 >$ KCl = $KNO_3 > K_2HPO_4 >$ no salt.

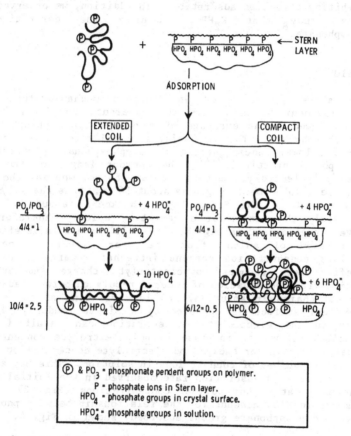

Figure 6. Schematic representation of the proposed phosphonylated polyelectrolyte adsorption mechanism. This schematic ignores the possibility of intercoil entanglements which probably occur at concentrations sufficiently elevated for the compact-coil mechanism to be operative.

Calcium enhancement beyond that explained by intrachain charge screening must be due to Ca^{++} binding, possibly by chelation with phosphonate groups within or between polymer chains. We observed a dramatic decrease in solution viscosity when less than equivalent amounts of Ca^{++} were added to polyphosphonate solutions, while observing no precipitation.[8] Thus it is quite likely that calcium binding reduces intrachain charge repulsions and forms crosslinks between chains. This mechanism accounts for both the observed reduction in solution viscosity and the enhancement of adsorption.

When compared to KCl, K_2HPO_4 is seen in Fig. 5 to actually have an inhibiting effect on adsorption. In addition, we observed in a previous study[8] that K_2HPO_4 enhances the desorption of polyphosphonate.

DISCUSSION

The above results support an adsorption mechanism that occurs via displacement of HPO_4 from the adherent, boundary, "Stern" layer, followed by an exchange of crystal-surface phosphate for polymer phosphonate. This mechanism is similar to those described earlier for lower concentrations of polyphosphonate electrolytes and for phosphoproteins.[16,17] The contour length of the fully extended polyelectrolyte chain is about 164 nm, whereas the root-mean-square displacement length is about 16 nm. Thus the polymer is large compared to the Stern layer thickness (about 0.5 nm[15], approximately the length of a repeating unit). Therefore the relative charge density along the chain will have a large effect on the repulsion of anions from the Stern layer. Increased electrolyte concentration screens intrachain charge repulsions. This effectively reduces polyelectrolyte charge and enhances adsorption via a combination of reduced repulsion between adsorbed-chains and chains in solution, and less crowding at the surface because less surface is occupied by a single random coil. An increase in the electrolyte concentration can result from HA dissolution. Also, an increase in polyelectrolyte concentration will result in a higher background electrolyte concentration due to the increased number of counter ions in solution. This may account for the increase in adsorption rate observed in the initial stages of adsorption at higher polymer concentrations (Fig. 2). These effects should also account for the relationship between phosphate released and phosphonate groups adsorbed, shown in Fig. 4.

Fig. 6 shows schematically how an extended coil conformation can be adsorbed and spread out along the surface, displacing a high proportion of phosphate ions into solution. Conversely, a collapsed or "compact" conformation is expected to occupy relatively few sites per coil, displace a low proportion of phosphate ions and form a relatively thick adsorption layer.

Calcium ion binding by polyphosphonate would enhance the collapse of the macromolecules and consequently promote their adsorption. Ca^{++} enhancement of adsorption may also explain the absence of calcium (as detected by atomic absorption) in solution during the adsorption experiments. The addition of phosphate ions (via K_2HPO_4) would lead to competition for the available sites on the hydroxyapatite surface. This is evident from the observed suppression of polyphosphonate adsorption when phosphate is added, compared to KCl addition (Fig. 5).

CONCLUSIONS

It may be concluded that the adsorption process depends upon the conformation of the random coil polyelectrolyte. This conformation is highly dependent upon the degree of intrachain charge repulsion and thus upon conditions which screen or neutralize the charges on the polymer chain. Also, divalent ion binding has a strong influence on polymer conformation. Thus, the adsorption of phosphonylated polyelectrolytes on hydroxyapatite may be described as follows: In solution, when a macromolecular chain containing phosphonic acid groups approaches an HA surface, its large concentration of negative charges repulses boundary-layer anions (phosphate in particular). This upsets the equilibrium between solution phase and solid phase phosphate. During the reestablishment of this equilibrium, some of the chain-attached phosphonate groups exchange with phosphate groups in the crystal lattice at the mineral surface, thereby binding the polymer to the surface. If the polymer chain is in an extended conformation it will lay down along the crystal face and bind at a relatively large number of sites, occupy a relatively large surface area and greatly hinder further adsorption. If the polymer concentration is low, adsorbed chains will tend to continue to spread out, thereby continuing to release more phosphate ions and to occupy more adsorption sites. As polymer concentration is increased, however, the total ionic strength increases and the random coil contracts. Additionally, as concentration increases entanglements will increase also. Both of these effects decrease the number of phosphonate groups per chain that are available to attach to the surface. The result is an increased number of solvated chain segments extending outward from the surface into the solution. The amount of polymer removed from solution per unit of surface area (e.g., adsorption) therefore increases with polymer concentration. This results in a thicker adsorption layer compared to when an extended coil undergoes adsorption from more dilute solutions. Conditions which promote the contracted-coil conformation are expected to result in the formation of the above type of relatively thick adsorption layer. Among such conditions are pH values well below 7, Ca^{++} binding, poor solvation, and high electrolytic strength. In the collapsed conformation adsorption occurs rapidly, and phosphate ions are released into solution over a very short time period, while the opposite occurs when the random coil is in an extended state. This can be attributed to the reduced effective charge and the smaller excluded volume of the contracted state, which reduces interchain repulsion and crowding at the HA surface.

After a contracted polymer coil has attached at several phosphate exchange sites, the loops and tails extending into the solution will probably entangle with other chains that are still in solution. This partially accounts for the slow increase in adsorption seen over longer times, but various other factors probably also play a role. Any calcium ions that are released during dissolution and adsorption must become bound by the adsorbed polyphosphonate. This would cause adsorbed polymers to contract still further, which would then allow further adsorption to take place. Ca^{++} bridging between chains probably also takes place and would result in further removal of polymer from solution via a second-layer effect. Eventually, Ca^{++} release by dissolution should also cause precipitation of some of the remaining polymer. This precipitated polymer may or may not become attached to hydroxyapatite in either the first or subsequent adsorption layers, but it is removed from solution in any case. In our experiments we can only observe the polymer remaining in solution and thus cannot directly distinguish precipitation from adsorption.

These observations should provide insights of importance when considering mechanisms involving acid demineralization, adhesion, protection against dental caries, site specific drug delivery to bone and teeth, flocculation, and other surface-controlled technical and biomedical phenomena.

ACKNOWLEDGEMENTS

        This work was carried out in the Laboratory for Materia Technica, Dental School, State University of Groningen, The Netherlands during the tenure of a Fogerty Senior International Fellowship (NIH grant TW00404). Other support was provided by an NIH/NIDR research grant (DE05596) and an NIH Career Development Award (DE00050). The polymer syntheses were carried out at the Gulf South Research Inst. in New Orleans, LA. Special thanks are due to Prof. J. Arends, Prof. J. ten Bosch, Dr. E. Moreno, and Dr. T. Bartels for the provision of laboratory facilities and for their encouragement and many helpful discussions. We also acknowledge the contributions of Mrs. B.F. Zimmerman, and the LSUSD Learning Resources Department and Word Processing Center, in preparation of this manuscript.

REFERENCES

1. M.D. Francis, The inhibition of calcium hydroxyapatite crystal growth by polyphosphonates and polyphosphates, Calc. Tiss. Res. 3:151-162 (1969).

2. A. Jung, S. Bisaz and H. Fleisch, The binding of pyrophosphate and two diphosphonates by hydroxyapatite, Calc. Tiss. Res. 11:269-280 (1973).

3. T. Bartels and J. Arends, Adsorption of a polyphosphonate on bovine enamel and hydroxyapatite, Caries Res. 13:218-226 (1979).

4. M. Anbar, G.A. St. John and T.E. Edward, Organic polyphosphonates as potential preventive agents for dental caries: in vivo experiments, J. Dent. Res. 53:1240-1244 (1974).

5. T. Bartels, J. Schuthof and J. Arends, The adsorption of two polyphosphonates on hydroxyapatite and their influence on the acid solubility of whole bovine enamel, J. Dent. 7(3):221-229 (1979).

6. M.D. Francis and R.L. Centner, The development of diphosphonates as significant health care products, J. Chem. Ed. 55: 760-766 (1978).

7. M. Anbar, G.A. St. John and A.C. Scott, Organic polymeric polyphosphonates as potential preventive agents of dental caries: in vitro experiments, J. Dent Res. 53:867-878 (1974).

8. H.R. Rawls, T. Bartels and J. Arends, Binding of polyphosphonates at the water/hydroxyapatite interface, J. Colloid Interface Sci. 87:339-345 (1982).

9. Y.S. Lipatov and L.M. Sergeeva, Structure of the adsorbed film and conformation of the adsorbed chain, in: "Adsorption of Polymers," D. Slutzkin, ed., Wiley & Sons, N.Y. (1974).

10. I. Cabasso, J. Jagur-Grodzinski and D. Vofsi, Synthesis and characterization of polymers with pendant phosphonate groups. J. Appl. Polym. Sci. 18:1969 (1974).

11. D.N. Fogg, and N.T. Wilkinson, The colorimetric determination of phosphorus, Analyst 83:406-417 (1952).

12. M. Kresak, E.C. Moreno, R.T. Zahradnik and D.I. Hay, Adsorption of amino acids onto hydroxyapatite, J. Colloid Interface Sci. 59(2):283-292 (1977).

13. J. Arends, Zeta potentials of enamel and apatites, J. Dentistry 7:246-253 (1979).

14. F. Oosawa, Characterization of polyelectrolytes, in: "Polyelectrolytes," Marcel Dekker, N.Y. (1971).

15. P. Sennett and J.P. Olivier, Colloidal dispersion, electrokinetic effects and the concept of zeta potential, in: "Chemistry and Physics of Interfaces," D.E. Gushee, ed., American Chemical Society, Wash., D.C. (1965).

16.   J.J. Klüppel and W. Plöger, The sorption of anions on
       synthetic apatite-kinetics and mechanism, Caries Res.
       15:188 (1981).
17.   A.C. Juriaanse, M. Booij, J. Arends and J.J. ten Bosch, The
       adsorption in vitro of purified salivary proteins on bovine
       dental enamel, Archs. Oral Biol. 26:91-96 (1981).

SURFACE CHEMICAL CHARACTERISTICS AND

ADSORPTION PROPERTIES OF APATITE

P. Somasundaran and Y.H.C. Wang

School of Engineering and Applied Science
Columbia University
New York, NY 10027

ABSTRACT

Interfacial behavior of apatites is governed to a large extent by their electrochemical properties which in turn are determined by pH, concentration of calcium, phosphate and fluoride. Adsorption of surfactants and polymers on apatite is dependent, among other factors, on the interfacial potential of the apatite. In this paper electrokinetic properties of synthetic hydroxyapatite and natural ore apatite containing fluoride are reported as a function of the pH, KNO_3, $Ca(NO_3)_2$, K_2HPO_4 and KF and mechanisms governing the surface charge generation are reviewed. Electrokinetic effects obtained for apatite upon treatment with concentrated KF solutions and calcite supernatant are analyzed to determine possible chemical alterations of its surface. Adsorption properties of ionic surfactants and ionic and nonionic polymers on apatite at different pH values are also discussed.

INTRODUCTION

Surface charge is an important property of a solid since it can determine as to what can adsorb, penetrate or adhere. Indeed, processes such as adsorption, particularly of surfactants or macromolecules, can alter the interfacial behavior of the solids markedly. While considerable information is available on surface charge characteristics of oxides such as alumina and silica, less is known about the behavior of sparingly soluble minerals such as apatite. Surface charge properties of apatite type materials are affected by many more variables and the mechanisms governing the charge generation are much more involved.

Properties of tooth apatite and its resistance to dental cavities have been known to be affected by the presence of chemical species of fluoride (1-4), vanadium (5,6), tin (7-9), molybdenum (10,11), chelating agents (12-14), and long-chain surfactants (15-18). The role of the electrochemical properties of the surface in determining the uptake of these chemical species by apatite is not, however, established. Furthermore, electrokinetic studies on apatites have produced results that are often in conflict with each other. Systematic work of Saleeb and de Bruyn (19) has yielded values for the point of zero charge of hydroxyapatite and fluorapatite that are in agreement with those obtained by us for natural ore apatite (20-23) and synthetic apatite during this work. Bell et al., (24) have also attempted to determine point of zero charge of apatites using titration techniques; this technique is, however, applicable strictly only for insoluble materials. Also, some of the prior work has been done with sodium salt solution as the supporting electrolyte. Since sodium can substitute for calcium in apatite lattice and thereby change its properties, the results obtained in sodium salt solution cannot be considered to truly represent that of apatite.

Surface charge of a mineral is determined by the concentration of "potential determining ions" in solution; the potential determining ions in the case of apatite can be the lattice ions or their reaction products with water. Other inorganic species, surfactants as well as polymeric reagents can also affect the interfacial charge and this in turn can be important in controlling transport of ions through the surface layers (25,26), but these secondary effects are controlled primarily by the potential determining ions. It is important to identify these ions since their concentrations do control the overall surface behavior. In this paper, our work to determine the role of various ionic species of Ca, PO_4, OH^-, and F^- and adsorption properties of selected surfactants and polymers of different charge characteristics are discussed.

EXPERIMENTAL

Both natural ore apatite $(Ca_{10}(PO_4)_6(F,OH)_2)$ and synthetic hydroxyapatite were used in this study. Zeta potential was measured using streaming potential technique (27) in the case of natural apatite and using electrophoresis in the case of synthetic hydroxyapatite. The method used for the study of the natural apatite essentially involved measuring streaming potential of 35/65 mesh particles (cleaned using dilute nitric acid) in potassium nitrate solutions adjusted to different pH values (20-22). In addition, the tests also included the pH measurement and the analysis of approximate fluoride content using a fluoride electrode. Aqueous solutions containing various amounts of calcium nitrate, potassium dihydrogen phosphate, and potassium fluoride were used to determine the role of calcium, phosphate, and fluoride ions.

Synthetic hydroxyapatite was prepared using precipitation technique by mixing appropriate amounts of K_2HPO_4 in KOH solution and
$Ca(NO_3)_2$ in water and boiling. Settled and dried crystals were
freeze dried and characterized using X-ray diffraction and chemical
analysis. Samples were stoichiometric (Ca/P = 1.67) and had hydroxyapatite structure. The zeta potential of the synthetic samples was
determined using Zeta-meter after the electrolyte solutions adjusted
to constant ionic strength values by adding sufficient KNO_3 and containing apatite were aged overnight and then equilibrated at the desired pH for 1 hour. Zeta potential and supernatant pH were measured
and filtrates were analyzed for calcium using titration or atomic
absorption spectrophotometer (EDTA was added to eliminate the interference of phosphate) and for total phosphate using a colorimetric
technique.

RESULTS AND DISCUSSION

Results obtained for zeta potential of synthetic hydroxyapatite
as a function of pH are given in Figure 1 and those obtained earlier
(21) for natural apatite in Figure 2. It can be seen that the synthetic hydroxyapatite has an isoelectric point of about 7 as opposed to
5.6 for the natural apatite and that H^+ and OH^- are potential deter-

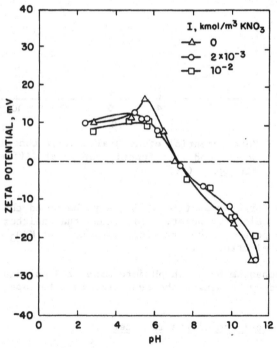

FIGURE 1.   Zeta potential of synthetic hydroxyapatite as a function
of pH at different ionic strengths.

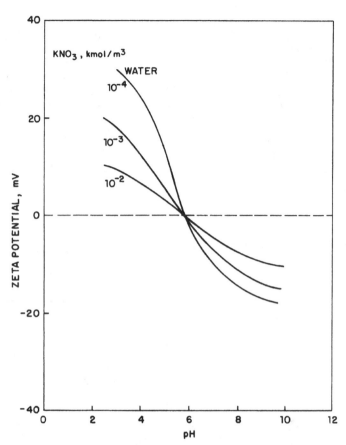

FIGURE 2.   Zeta potential of natural apatite containing fluoride
             determined as a function of pH at different ionic
             strengths (21)

mining ions for both apatites. It is to be noted that KNO_3 does not
shift the isoelectric point. This means that neither K^+ nor NO_3^- has
any surface potential determining role and that KNO_3 can be used as
a reference electrolyte.

The mechanism by which pH determines the surface potential can
be developed by examining the reactions that the apatite can undergo in
water.

Apatite Chemical Equilibria in Water

(1) $Ca^{2+} + OH^- \rightleftarrows CaOH^+$ $10^{1.4}$ (28)

(2) $CaOH^+ + OH^- \rightleftarrows Ca(OH)_2(aq)$ $10^{1.37}$ (29)

(3) $Ca(OH)_2(ag) \rightleftarrows Ca(OH)_2(s)$ $10^{2.45}$ (29)

(4) $H_3PO_4 \rightleftarrows H^+ + H_2PO_4^-$ $10^{-2.15}$ (30)

(5) $H_2PO_4^- \rightleftarrows H^+ + HPO_4^{2-}$ $10^{-7.2}$ (30)

(6) $HPO_4^{2-} \rightleftarrows H^+ + PO_4^{3-}$ $10^{-12.3}$ (30)

(7) $Ca^{2+} + HPO_4^{2-} \rightleftarrows CaHPO_4(aq)$ $10^{2.7}$ (32)

(8) $CaHPO_4(aq) \rightleftarrows CaHPO_4(s)$ $10^{4.3*}$

(9) $Ca^{2+} + H_2PO_4^- \rightleftarrows H_2PO_4^- \rightleftarrows CaH_2PO_4^+(aq)$ $10^{1.08}$ (32)

When the pH is increased, equations 1 through 6 will be driven towards
the right hand side and this will result in a reduction in the
activities of the positive species and an increase in those of the
negative species. The net result of these reactions will be an excess
of negative potential determining ions in solution and this will make
the surface negatively charged. When the pH is decreased, the above
reactions will move in the reverse direction and the surface will
become positively charged. If the above reactions are indeed responsible for the surface charge generation, then for apatite, which can be
represented as $M_{10}(PO_4)_6Z_2$ where M can be calcium, barium, strontium
etc. and Z stands for hydroxyl, fluoride etc., addition of any of the
species should change the surface potential in a manner dictated by
the following equations:

(10) $M_{10}(PO_4)_6 Z_2 \rightleftarrows 10M^{2+} + 6PO_4^{3-} + 2Z^-$

(11) $K_{sp} = (M^{2+})^{10} (PO_4^{3-})^6 (Z^-)^2$

$M^{2+} = Ca^{2+}$ $Z^- = OH^-$

(12) $\psi_o = \dfrac{RT}{2F} \ln \dfrac{(Ca^{2+})}{(Ca^{2+})_{pzc}}$

---

*Calculated from equilibrium constants for reaction [7] and $CaHPO_4(s)$
$\rightleftarrows Ca^{2+} + HPO_4^{2-}$, $K_{sp} = 10^{-7}$.

$$(13) \quad \psi_o = - \frac{RT}{F} \ln \frac{(OH^-)}{(OH^-)_{pzc}}$$

$$(14) \quad \psi_o = - \frac{RT}{3F} \ln \frac{(PO_4^{3-})}{(PO_4^{3-})_{pzc}}$$

It is to be noted at this point that since more than two types of species can determine the surface potential of hydroxyapatite, there is not just one point of zero charge as in the case of silica or alumina but many points of zero charge that will constitute a line of zero charge on a calcium-phosphate-hydroxyl ternary diagram. A zero zeta potential curve can be constructed on such a diagram if the condition of constant solubility product is met.

The role of calcium and phosphate as potential determining species for apatite can be tested by conducting tests as a function of their concentrations below as well as above its isoelectric point. If a cation is potential determining, then its addition will make it more positively charged above and below the isoelectric point, that is, whether the mineral is originally negatively charged or positively charged. Similarly an anion, if potential determining, will make it more negatively charged both above and below the isoelectric point. Strictly, variation of the surface potential itself with respect to the potential determining ions should be dictated by the Nernst relationship. In a system where the mineral dissolves and the resultant species undergo a complex set of step-wise reactions, it is, however, difficult to estimate either the activity of relevant species or the surface potential. The indirect test on the effect of species under conditions when the mineral is similarly charged is more useful for such systems. The effect of phosphate, calcium and fluoride at pH 10 (above the isoelectric point) is illustrated in Figure 3. Note that the ionic strength is kept constant. It can be seen that calcium, as expected, makes the mineral less negative and then even reverses the charge to make it positive at pH 10 with 10^{-4} kmol/m^3 calcium salt addition. Importantly, phosphate makes the mineral more negatively charged even though the mineral is already negative; this confirms the potential determining role of the phosphate. It might be noted that fluoride does not produce any significant effect under these conditions. The effect of these species at pH 5 (below the isoelectric point) is illustrated in Figure 4. Note that calcium has an effect under these conditions also, but now only at 4×10^{-3} kmol/m^3 and above essentially because of the presence of significant amounts of dissolved calcium (3×10^{-3} kmol/m^3). While phosphate has the expected effect to make the mineral more negative at this pH, flouride, an anion is observed to make the mineral more positive.

The above species can exist in the form of various complexes with each other as well as with H$^+$ or OH$^-$. The effect on surface

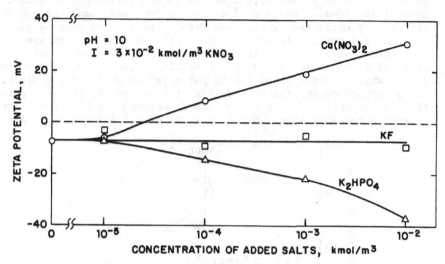

FIGURE 3. Zeta potential of synthetic hydroxyapatite determined as a function of concentration of $Ca(NO_3)_2$, KF and K_2HPO_4 at pH 10.

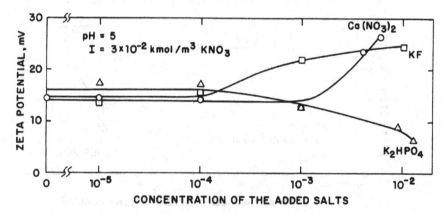

FIGURE 4. Zeta potential of synthetic hydroxyapatite determined as a function of concentration of $Ca(NO_3)_2$, KF and K_2HPO_4 at pH 5.

potential can be either direct or due to complexation with potential determining ions. The role of different ions and complexes can be more easily estimated by plotting zeta potential in various electrolytes as a function of total concentration of the relevant species. Zeta potential is plotted in this manner as a function of total P and Ca at pH 5 and 10 in Figures 5 to 8. It is seen from Figures 5 and 7 that while calcium changes zeta potential at pH 5 <u>without</u> altering the phosphate concentration significantly, at pH 10, the zeta potential change upon addition of calcium is accompanied by a reduction in phosphate concentration. Higher effects of calcium in the alkaline pH

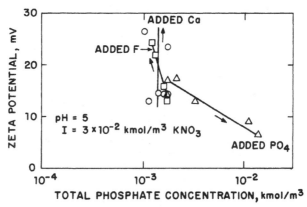

FIGURE 5.   Zeta potential of synthetic hydroxyapatite in $Ca(NO_3)_2$, K_2HPO_4 and KF solutions at pH 5 as a function of total phosphate concentration. Ca, PO_4 and F additions correspond to the concentrations shown in Figure 4.

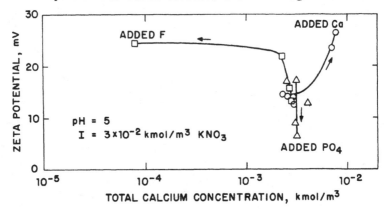

FIGURE 6.   Zeta potential of synthetic hydroxyapatite in $Ca(NO_3)_2$, K_2HPO_4 and KF solutions at pH 5 as a function of total calcium concentration. Ca, PO_4 and F additions correspond to the concentrations in Figure 4.

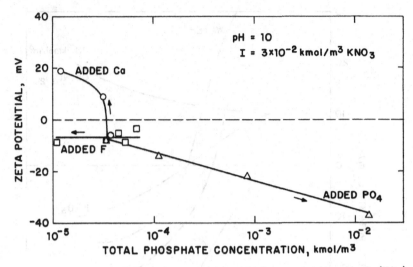

FIGURE 7.   Zeta potential of synthetic hydroxyapatite in $Ca(NO_3)_2$,
K_2HPO_4 and KF solutions at pH 10 as a function of total
phosphate concentration. Ca, PO_4 and F additions correspond to the concentrations in Figure 3.

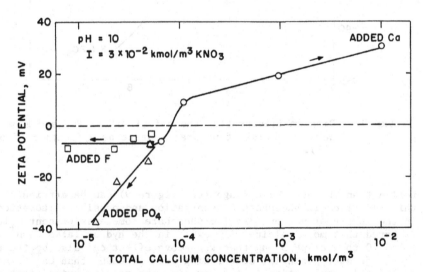

FIGURE 8.   Zeta potential of synthetic hydroxyapatite in $Ca(NO_3)_2$,
K_2HPO_4 and KF solutions at pH 10 as a function of
total calcium concentration. Ca, PO_4 and F additions
correspond to the concentrations in Figure 3

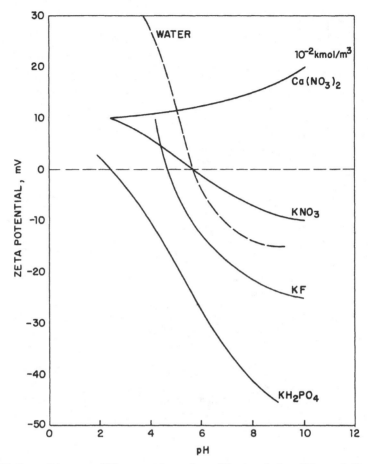

FIGURE 9.   Diagram illustrating the effect of Ca, PO_4 and F on the
zeta potential of natural apatite as a function of pH
(21).

region than in the acidic range (see Figure 9) can be attributed to
this reduction in phosphate concentration and possible alterations
in the levels of various calcium phosphate complexes present. It
is noted that more calcium will exist in the hydroxylated $CaOH^+$ form
at pH 10 than at pH 5 and that the above effect can also be the result
of the greater potential determining role of $CaOH^+$ than Ca^{++}. An
additional complication in interpreting the results arises from the
presence of a larger amount of calcium in acidic solutions, thus
reducing the effect of added calcium (see Table 1). From Figures
6 and 8 it is seen that phosphate also acts in the same manner directly at pH 5 and with some reduction in calcium concentration at pH 10.

Table 1. Total Ca and P Concentrations in Some of the Test Solutions

Solution	pH	Ca conc. $(kmol/m^3)$	P conc. $(kmol/m^3)$
(I) Triple distilled	11.2	3.75×10^{-6}	$< 1.6 \times 10^{-5}$
water	10.3	$< 2.5 \times 10^{-5}$	$< 1.6 \times 10^{-5}$
	7.1	1.5×10^{-5}	$< 1.6 \times 10^{-5}$
	6.2	1.24×10^{-4}	1.13×10^{-4}
	5.57	1.17×10^{-3}	4.87×10^{-4}
	4.85	4.41×10^{-3}	2.7×10^{-3}
	4.7	5.12×10^{-3}	3.29×10^{-3}
	2.9	5.62×10^{-3}	3.35×10^{-3}
(II) $2 \times 10^{-3} kmol/m^3$	11.35	$< 2.5 \times 10^{-5}$	$< 1.6 \times 10^{-5}$
KNO_3	10.3	2.75×10^{-5}	$< 1.6 \times 10^{-5}$
	7.4	4.38×10^{-5}	$< 1.6 \times 10^{-5}$
	5	3.68×10^{-3}	2.73×10^{-3}
	2.5	5.51×10^{-3}	3.48×10^{-3}
(III) $10^{-2} kmol/m^3$	11.2	4.75×10^{-6}	$< 1.6 \times 10^{-5}$
KNO_3	9.85	3.38×10^{-4}	$< 1.6 \times 10^{-5}$
	7.8	1.27×10^{-4}	8.06×10^{-5}
	5.8	4.2×10^{-3}	2.69×10^{-3}

Fluoride is seen to produce a marked reduction in calcium concentration at all pH values. The effect of fluoride addition on zeta potential is also peculiar in that it makes the mineral more positive in the acidic region and slightly more negative in the alkaline region. This can be clearly seen in Figure 9. This information can also be used to determine the surface chemical effects of fluoride. It has been suggested that fluoride in sufficient amounts can form fluorapatite at high pH values and fluorite at low pH values. Fluorite is reported to show a higher point of zero charge than hydroxyapatite (32,33) and fluorapatite a lower point of zero charge (19). Therefore, it requires the formation of fluorite on the surface to make the mineral more positively charged and the formation of fluorapatite to make it more negatively charged. The zeta potential changes observed here in the present study with the addition of fluoride suggests the formation of fluorite at the surface of hydroxyapatite at low pH values and fluorapatite at high pH values. Solution conditions also dictate the above possibility since the calcium concentration at pH 5 is about 3×10^{-3} kmol/m³ and CaF_2 with a solubility product of 3.95×10^{-11} can be expected to form above about 10^{-4} kmol/m³ fluoride (34). Formation of CaF_2 is also indicated by the observed decrease in calcium concentration at this pH from 3×10^{-3} to 8×10^{-6} kmol/m^3 in 10^{-2} kmol/m^3 fluoride solution. At pH 10 also, with a total calcium concentration of 4×10^{-5} kmol/m^3, fluorite can precipitate above about 10^{-3} kmol/m^3 fluoride concentration. In the present case, however, more of the calcium will be present in complexed forms with hydroxyl and also with phosphate since the stability of CaHPO$_4$ (the predominant species at pH 10) is higher than that of Ca$_2$HPO$_4^+$ (the predominant species at pH 5). Precipitation of fluorite will therefore require larger concentrations of fluoride at this pH.

In view of the above findings, the permanent nature of the above effects was investigated by soaking natural apatite in 1 kmol/m^3 KF solution at neutral pH and then determining the zeta potential after washing for almost 8 hours (35). The isoelectric point of the natural apatite was found to shift from 5.5 to about 7.6 (Figure 10). It is indeed likely that if the measurements had been done after rinsing it for a few minutes instead of 8 hours, a larger shift of the iso-electric point would have resulted. In order to determine the

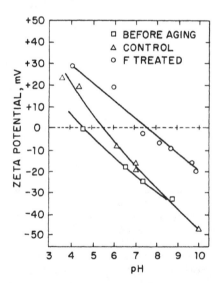

FIGURE 10. Diagram illustrating the effect of prolonged fluoride contact on the electrochemical properties of natural apatite. Zeta potentials in 2×10^{-3} kmol/m^3 KNO$_3$ solution of freshly cleaned apatite, cleaned and aged apatite, and F-treated (using 1 kmol/m^3 KF following cleaning and aging) apatite.

permanence of this effect, zeta potential of three fluoride treated
samples at initial pH values of 4, 7 and 10 were measured as a function of time.

Also pH, and F concentrations of the supernatants were measured.
Figure 11 shows the change in the parameters at pH 7. It is noted
that the change in the zeta potential of fluoride treated sample,
4 mV, is about one-third of the untreated sample. When the simultaneous change in pH is taken into account, it becomes apparent that
even the 4 mV change is that which is expected with the observed
change in pH. It is interesting that the fluoride concentration of
fluoride treated system did increase while that of the solution containing the untreated sample did not change significantly. Results
obtained at pH 4 and pH 10 and given in Figures 12 and 13 also show the same
trend except for the slight increase observed for the untreated sample
at pH 4. Treatment with concentrated fluoride solutions (such as
1 kmol/m^3 KF solution used) is thus found to make the natural apatite
sample more positively charged and interfacial potential values more
stable, again supporting the possibility of fluorite precipitation on
the apatite surface during contact with sufficiently concentrated
fluoride solutions. Implications of these changes on the solubility
of the mineral as well as the rate of transport of charged chemical
species into apatite are to be recognized.

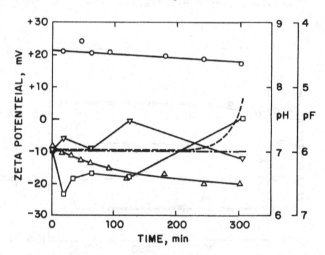

FIGURE 11.   Changes in zeta potential, F concentration in solution
and pH with time for fluoride treated and untreated
(control) natural apatite/potassium nitrate systems in
the neutral pH region; O zeta potential-F treated,
△ zeta potential-untreated, ☐ pH-F treated, ▽ pH-
untreated, ----F - F treated, -·-·- F - untreated.

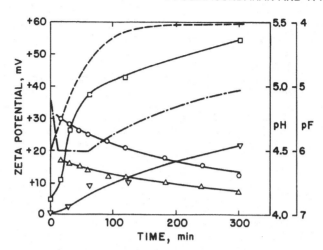

FIGURE 12. Changes in the zeta potential, F concentration in solution, and pH with time for fluoride treated and untreated
natural apatite/potassium nitrate systems in the acidic
pH range. O zeta potential - F treated, △ zeta
potential - untreated, ☐ pH - F treated, ▽ pH - untreated, ---- F - F treated, --·-;- F - untreated.

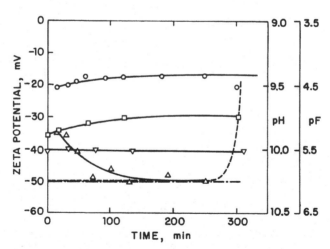

FIGURE 13. Changes in the zeta potential F concentration in solution,
and pH with time for fluoride treated and untreated
natural apatite/potassium nitrate systems in the alkaline pH range. O zeta potential - F treated, △ zeta
potential - untreated, ☐ pH - F treated, ▽ pH -
untreated, --- F - F treated, --·--· F - untreated.

It is important to note that the electrochemical properties of
the apatite are very much dependent upon the method of preparation
of the mineral, cleaning, storing etc. (22). Also other minerals
present in the system can alter the surface properties totally. For
example, conditioning of synthetic hydroxyapatite in calcite supernatant (obtained by stirring calcite in water and removing all the
solid particles by centrifugation) is found to shift the isoelectric
point to that of the calcite suggesting precipitation of calcite or
other possible solids on the hydroxyapatite surface (Figure 14,(36)).
Solid-solution equilibria in mineral systems containing apatite can
be extremely complex, leading to the possibility of precipitation of
different minerals depending upon the solution conditions such as
pH (37-40). The results suggest that in real systems containing
more than one mineral, behavior of the minerals can be totally
different from their behavior when present alone.

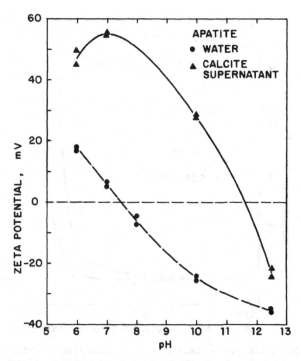

FIGURE 14.   Diagram illustrating the effect of calcite supernatant
              on the zeta potential of synthetic hydroxyapatite (36).

As mentioned earlier, surface charge characteristics are extremely important in determining adsorption of different species on
hydroxyapatite. This is found to be the case with anionic surfactant, dodecylsulfonate (Figures 15 and 16). While adsorption of the a-
nionic sulfonate is significant at pH 6.7, there is no adsorption at pH
10.7. In contrast to the sulfonate, the dodecylamine is found to
adsorb at both pH values. While zeta potential results of Mishra
et al.(39), also suggest adsorption of dodecyltrimethylammonium
chloride and dodecylammonium chloride under all pH conditions at
a concentration of 10^{-3} kmol/m^3, at lower concentrations of
dodecylamine significant shift in zeta potential is observed only
above the isoelectric point. Dodecylsulfonate, on the other hand,
lowers the zeta potential at all levels in the reported pH range
5 to 10. Oleate also lowers the zeta potential, shifting it towards
that of Ca-oleate at high oleate to apatite ratios. Evidently, the
precipitation of Ca-surfactant salt can also be a major phenomenon
in these systems.

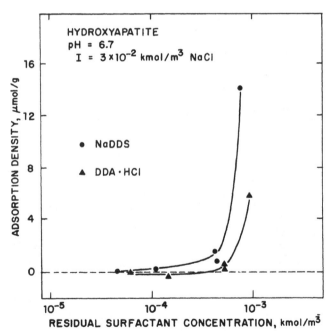

FIGURE 15.   Adsorption isotherms of dodecylsulfonate and dodecylamine
            on synthetic hydroxyapatite at pH 6.7 at an ionic strength
            of 3 x 10^{-2} kmol/m^3 NaCl.

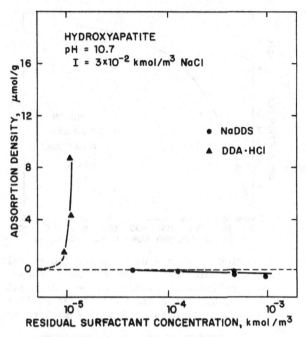

FIGURE 16.   Adsorption isotherms of dodecylsulfonate and dodecylamine
             on synthetic hydroxyapatite at pH 10.7 at an ionic
             strength of 3 x 10^{-2} kmol/m^3 NaCl.

     Adsorption of polymers on hydroxyapatite are also found to
be influenced by their charge characteristics. Adsorption of selected
nonionic, anionic and cationic polymers on hydroxyapatite at pH 11
and 6.6 is shown in Figures 17 and 18 respectively. While in the
alkaline pH range where the mineral is negatively charged, only the
cationic polymer is found to adsorb, in the neutral pH range all the
polymers are found to adsorb. Polymer adsorption on solids is considered to result mainly from hydrogen bonding, electrostatic bonding
and covalent bonding depending on the mineral/polymer system (41).
Evidently hydrogen bonding might be sufficiently active to cause
adsorption of all the polymers at neutral pH values. It is noted
that there is no measurable adsorption of even the nonionic polymer at pH 11.1. This is attributed to the hydrolysis of the polymer to the anionic form in alkaline solutions (42) and the electrostatic repulsion between the resultant functional group on the polymeric species and similarly charged mineral particles. It is evident that adsorption of macromolecules will be a complex function of
not only the properties of the solid and the polymer but also the

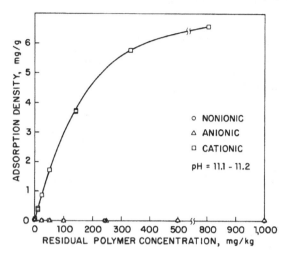

FIGURE 17.   Adsorption isotherms of nonionic (polyacrylamide) anionic
             (polyacrylamide containing carboxyl group) and cationic
             (polyacrylamide containing amine functional group) poly-
             mers on synthetic hydroxyapatite at pH 11.1 to 11.2
             at an ionic strength of 3 x 10^{-2} kmol/m^3 NaCl.

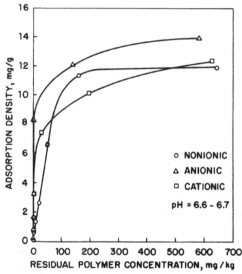

FIGURE 18.   Adsorption isotherms of nonionic (polyacrylamide), anionic
             (polyacrylamide containing carboxyl group) and cationic
             (polyacrylamide containing amine group) polymers on
             synthetic hydroxyapatite at pH 6.6 - 6.7 at an ionic
             strength of 3 x 10^{-2} kmol/m^3 NaCl.

solution conditions and possible alterations of both the solid and
the polymer in the solutions. It is important to note that adsorption of the polymers on solid can also depend on that of the presence
of any surfactant in solution (43). Other interfacial properties of
the mineral such as wettability can also be affected markedly by
such polymer-surfactant interactions (49).

CONCLUSIONS

Electrochemical property of apatite is a complex function of not
only pH, but also concentration of various constituent species as
well as the type of pretreatment of the solid. Hydrogen, hydroxyl,
calcium and phosphate play a potential determining role either
directly or by altering the concentration of other potential determining ions. While phosphate addition makes the mineral more negatively charged and calcium makes it more positively charged under
all pH conditions, fluoride is found to make the mineral more
positively charged in acidic solutions and more negatively charged
in alkaline solutions possibly due to the formation of fluorite
and fluorapatite respectively. Prolonged contact with fluoride also
does produce a significant increase in the isoelectric point suggesting some fluorite precipitation on the surface. Similarly, contact
of hydroxyapatite with calcite supernatant produced a shift of the
isoelectric point towards that of calcite. It is clear that in
apatite systems containing other minerals alterations in surface composition due to various precipitations are possible depending on the
mineral-solution equilibria at various pH values. The resultant
electrochemical nature of the mineral particles is found to play a
governing role in determining the adsorption of surfactants and polymers of various charge characteristics.

ACKNOWLEDGMENTS

The authors wish to acknowledge the support of the National
Institute of Health (5-R01-DE 03460) and the Chemical and Process
Engineering Division of the National Science Foundation. We also
thank S. Phillips, R.D. Kulkarni, and K.P. Ananthapadmanabhan for
technical assistance and discussion.

REFERENCES

1. J.C. Eliot, Calc. Tiss. Res., 3:293 (1969).
2. T.W. Cutress, Archs. Oral. Biol., 11:121 (1966).
3. J.A. Gray, J. Dent. Res., 44:493 (1965).
4. J.C. Muhler, T.M. Boyd and G. Van Huysen, J. Dent. Res., 29:182,
    (1950).

5. H.T. Dean, P. Jay and F.A. Arnold, F.J. McClure and E. Elvove,
   Public Health Reports, 54:862 (1939); B.F. Miller, Proc. Soc.
   Exp. Biol. and Med., 39:389 (1938); H.C. Hodge and S.B. Finn,
   Prac. Soc. Exp. Biol. and Med., 42:318 (1939).
6. C.F. Geyer, J. Dent. Res., 32:590 (1953).
7. G. Tank and C.A. Storuik, J. Dent. Res., 39:473 (1960).
8. D.M. Hadjimarkos, J. Pediat., 48:195 (1956).
9. W. Buttner, J. Dent. Res., 42:453 (1963).
10. T.G. Ludwig, W.B. Healy and F.L. Losse, Nature, 186:695 (1960).
11. G.N. Jenkins, Dr. Dent. J., 122:435-441; 500-503; 545-550 (1967).
12. L.H. Eggers, N.Y. St. Dent. J., 27:75 (1961).
13. A. Schatz and J.J. Martin, J. Am. Dent. Assoc., 65:368 (1962).
14. N.W. Johnson, Archs Oral Biol., 11:1421 (1966).
15. R.S. Manly and K.F. Manly, J. Dent. Res., 42:565 (1963).
16. H.M. Myers, J. Dent. Res., 42:1547 (1963).
17. A.H. Meckel, Arch. Oral Biol., 10:585 (1965).
18. T.J. Roseman, W.L. Higuchi, B. Hodes and J.J. Herferren, J. Dent.
   Res., 48(4):509 (1969).
19. F.Z. Saleeb and P.L. de Bruyn, Electroanal. Chem. and Interfac.
   Electrochem., 37:99 (1972).
20. P. Somasundaran, J. Coll. Interf. Sci., 27:659 (1968).
21. P. Somasundaran and G.E. Agar, Trans. AIME, 252:348 (1972).
22. P. Somasundaran, in "Clean Surfaces, Their Preparation and
   Characterization for Interfacial Studies", Marcel Dekker, N.Y.
   (1970).
23. Y.H. Wang, Zeta Potential Studies on Hydroxyapatite, M.S. Thesis,
   Columbia University (1975).
24. L.C. Bell, A.M. Posner and J.P. Quirk, Nature, 239:515 (1972);
   J. Coll. Interf. Sci. 42:250 (1973).
25. K. Sollner, J. Dent. Res., 53:266 (1974).
26. See J. Dent. Res., 53:308 (1974).
27. P. Somasundaran and R.D. Kulkarni, J. Coll. Interf. Sci.,
   45:591 (1973).
28. R.P. Bell and J.H.B. George, Transactions, Faraday Society,
   46:619 (1953).
29. D.D. Hedberg, "Sargent Chart of Equilibrium Contents of Inorganic
   Compounds", E.H. Sargent and Co., Chicago (1963).
30. C.D. Hodgman, R.C. Weast and S.M. Selby, "Handbook of Chemistry
   and Physics", 42nd ed., The Chemical Rubber Publishing Co. (1961)
31. J. Bjerrum, G. Schwarzenbach and L.G. Sillen, "Stability Constants of Metal Ion Complexes, with Solubility Products of Inorganic Substances", Part II, Special Publication No. 7, The
   Chemical Society, London (1958).
32. P. Ney, "Zeta-Potentiale und Flotierbarkeit Von Minevalen,
   Springer-Verlag", New York, (1973).
33. J.D. Miller and J.B. Hiskey, J. Coll. Interf. Sci., 41:567
   (1972).

34. N.A. Lang, ed. "Handbook of Chemistry", 10th ed., McGraw Hill, New York (1961).

35. S. Phillips, R.D. Kulkarni and P. Somasundaran, Effects of Pretreatment with Fluoride Solutions on Apatite Electrochemical Properties, Annual Meeting of American Association of Dental Research, 1975; J. Dent. Res., Feb.,1975, p. 180.

36. O.J. Amankonah and P. Somasundaran, unpublished results.

37. G.H. Nancollas, Z. Amjad and P. Koutsoukas, Calcium Phosphates-Speciation, Solubility and Kinetic Considerations, in "Chemical Modelling in Aqueous Systems", J.A. Jenne, ed., ACS, Washington, D.C. (1979).

38. Y. Avnimelech, E.C. Moreno and W.E. Brown, Solubility and Surface Properties of Finely Divided Hydroxyapatite, J. Res. NBS, 77A(1):149 (1973).

39. R.K. Mishra, S. Chander and D.W. Fuerstenau, Colloids and Surfaces, 1:105 (1980).

40. S. Chander and D.W. Fuerstenau, Colloids and Surfaces, 4:101 (1982).

41. G.C. Sresty, A. Raja and P. Somasundaran, in "Recent Developments in Separation Science", Vol. 4, CRC Press, West Palm Beach, FL (1978).

42. A.F. Hollander, P. Somasundaran and C.C. Gryte, in "Adsorption from Aqueous Solution", P. Tewari, ed., Plenum Press, New York (1981).

43. B.M. Moudgil and P. Somasundaran, Adsorption of Charged and Uncharged Polyacrylamides on Hematite, SME Preprint 82-160.

44. P. Somasundaran and L.T. Lee, Separation Science and Technology, 16:1475 (1981).

NEW NMR METHODS FOR THE STUDY OF HYDROXYAPATITE SURFACES

James P. Yesinowski,* Rex A. Wolfgang, and
Michael J. Mobley+

Miami Valley Laboratories and +Sharon Woods Technical
Center *The Procter & Gamble Co.
Cincinnati, Ohio 45247

INTRODUCTION

Improved surface-characterization techniques are needed to study the adsorption of molecules and ions from aqueous solutions onto microcrystals of the biological mineral hydroxyapatite, the prime constituent of bone and teeth. The continuing development of techniques for obtaining high-resolution nuclear magnetic resonance (NMR) spectra from solids indicates that NMR could provide a valuable spectroscopic characterization of hydroxy-apatite surfaces. We report here the successful application of new NMR techniques to two areas: (1) the adsorption onto the surface of hydroxyapatite of diphosphonates, used both as inhibitors of biological mineralization and as bone-scanning agents; (2) the reactions of hydroxyapatite with fluoride ion, which are important in the anti-caries benefits provided through fluoridation of dental enamel.

Before an introductory discussion of the two approaches we have used, it is worthwhile to consider the two severe experimental difficulties encountered with the use of NMR to study solid surfaces: limited sensitivity and low resolution. The problem of limited sensitivity can be alleviated by: (1) the use of powders having high specific surface areas; (2) the choice of a high-sensitivity nucleus for NMR observation, preferably ·one which is present only in the adsorbate, not the substrate; (3) the use of high-field pulsed Fourier-transform NMR spectro-meters with superconducting magnets ("supercons").

The primary reason for the low resolution seen for surface-adsorbed species is the solid-like nature of the adsorbate. A

151

nuclear spin in a typical solid experiences strong anisotropic
interactions, such as dipolar coupling to other nuclei and an
anisotropy in the chemical shift, which greatly broaden its NMR
absorption.   Although low-resolution NMR can often provide
information about the <u>motion</u> of adsorbed species, spectra are
more difficult to obtain than in the high-resolution case, and
chemical detail is usually lost.   Fortunately, fundamental
theoretical and experimental advances have been made in solid-
state NMR to develop techniques for obtaining "high-resolution
NMR of solids".   A description of some of these techniques and
their application to problems in surface science during the past
decade is given in a useful recent review.[1]

Figure 1 presents a schematic view of the problem of low
resolution in the NMR of solids, and of the two approaches we
have used to overcome this problem.   The first method we have
developed   utilizes   aqueous   colloidal   suspensions   of
hydroxyapatite to narrow the ^{31}P NMR peaks both of bulk solid

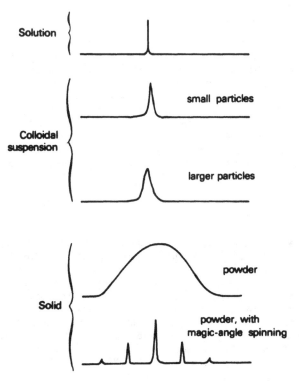

Fig. 1.   Schematic   NMR   spectra   illustrating   techniques   for
obtaining high-resolution NMR spectra of solids (e.g.
^{31}P NMR of calcium phosphates).

and of surface-adsorbed diphosphonates.[2]   This novel approach to high-resolution NMR of solids has some limitations which render it impractical for studying the surface reactions of fluoride ion with hydroxyapatite. We have therefore applied the method of [19]F "magic-angle spinning" (MAS) NMR to characterize, for the first time, these latter surface reactions.

We begin by discussing in some detail the ^{19}F MAS-NMR spectra of bulk fluorapatite, fluorohydroxyapatite solid solutions, and calcium fluoride,[3] which represent model compounds for the possible forms of fluoride on the hydroxyapatite surface. Differences in the ^{19}F NMR characteristics of these compounds have been exploited to analyze for the presence of these compounds in hydroxyapatite samples exposed to aqueous solutions of fluoride.[4]   It is already clear at this early stage that ^{19}F MAS-NMR is a powerful and sensitive technique for spectroscopically identifying the form of fluoride at the hydroxyapatite surface.

EXPERIMENTAL

     High-resolution ^{31}P NMR spectra (using 20 mm tubes) and ^{19}F MAS-NMR spectra were obtained on a Bruker CXP-300 spectrometer operating at 7.0T. The preparation of the colloidal hydroxyapatite suspension has been described previously.[2]

     The ^{19}F MAS-NMR probe was a modified Bruker double-tuned MAS probe, using Andrew-Beams rotors containing 0.3 cc. (~250 mg) of packed dry powder. The proton channel was retuned to the fluorine frequency, and all Teflon in the probe (including coaxial cables) was removed or replaced to eliminate a strong fluorine background signal. The high-power ^{19}F transmitter (~150-600w) resulted in a 90° pulse of typically 3-6 µs. The normal spinning speed of the Delrin rotors was 3.8 kHz and was measured either with a photo-diode tachometer or from the frequency separation of spinning sidebands in the spectra.

     Fluorapatite was synthesized by a literature method.[5] The fluorohydroxyapatite solid solutions were generously loaned by Dr. E. C. Moreno, Forsyth Dental Center, who provided information on the chemical analysis. Infrared spectra of these samples establishing the solid-solution character of the samples were obtained by Mr. Al Fehl at our Miami Valley Laboratories. Dr. Michael Munowitz, of the Department of Chemistry, Harvard University, and the Francis Bitter National Magnet Laboratory, very kindly obtained the theoretical MAS-NMR spectra using his simulation program.

     The hydroxyapatite sample treated with fluoride solutions

was a commercial Bio-Rad hydroxyapatite with a Ca/P molar ratio of 1.65 and a specific surface area measured by BET of 61.6 m^2/g. This was chosen for our initial studies because a high surface area improves the sensitivity of the NMR experiment. Fluoride treatment solutions were prepared with various concentrations of NaF and a NaCl concentration of 2.5×10^{-2}M to elevate the ionic strength. Hydroxyapatite powder was added as a slurry (0.1 g/ml) to a specific volume of fluoride treatment solution at 37°C and pH 7.0. For the samples whose ^{19}F NMR spectra are given in this paper, 10 ml. of slurry was added to 1 liter of solution, the pH was allowed to rise during the reaction, and the samples were not stirred. Further experiments using stirred samples, extensively washed, pre-equilibrated, and pH-statted, have yielded similar results. To measure the concentration dependence of fluoride uptake, 5 ml of slurry was added to 150 ml of fluoride solution (below 100 ppm F^-) or to 50 ml of fluoride solution (above 100 ppm F^-) and the samples were stirred. Fluoride treatments were for 1 hour. Crystals were isolated by filtering with a 0.22μ Millipore filter and were dried at 110°C for 1 hr. The amount of fluoride in the solid was determined by fluoride electrode measurements of the dissolved solids.

NMR OF COLLOIDAL SUSPENSIONS OF HYDROXYAPATITE

The ^{31}P NMR spectrum at 36 MHz of powdered hydroxyapatite, $Ca_{10}(OH)_2(PO_4)_6$, is a broad peak Gaussian in shape, with a half-height linewidth of 1.85 kHz. The peak is broad because of the presence of dipolar couplings between spins, both ^{31}P-^{1}H and ^{31}P-^{31}P, as well as a small anisotropy in the ^{31}P chemical shift.[6] In solution, these anisotropic interactions of the ^{31}P nuclear spin would be averaged to zero by the fast rotational tumbling and translational diffusion due to Brownian motion. The result would be a high-resolution NMR spectrum, with a sharp peak at the so-called "isotropic" chemical shift position. However, simply dissolving the sample is not a useful approach when we are interested in the intrinsic characteristics of the solid phase.

Several years ago, one of us (J.P.Y.) showed that the ^{31}P NMR peak of calcium hydroxyapatite could be narrowed by preparing a colloidal suspension of the solid in water.[2] The rotational correlation time characterizing the tumbling of the small (125x125x500Å) particles was small enough to eliminate the anisotropic interactions. Formation of a stable colloidal suspension required both the addition of a diphosphonate such as EHDP, $CH_3C(OH)(PO_3)_2^{-4}$, as a peptizing agent as well as prolonged ultrasonication to disaggregate the particles. The ^{31}P NMR spectrum at 36 MHz of the resultant concentrated

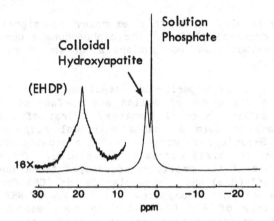

Fig. 2.   121.5 MHz ^{31}P NMR spectrum of colloidal suspension of
          1.15g   hydroxyapatite/0.1g $Na_2CH_3C(OH)(PO_3H)_2$ in
          10 ml. D_2O; 2w ^{1}H decoupling and a line-broadening
          of 5 Hz were used; see text and reference (2) for
          further details.

colloidal suspension showed a thirty-fold reduction in the
hydroxyapatite linewidth to a Lorentzian peak with a half-height
width of 72Hz, in reasonable agreement with theoretical
predictions.

    Furthermore, the ^{31}P NMR spectrum showed a peak arising
from the diphosphonate adsorbed on the surface of the
hydroxyapatite. Figure 2 shows a ^{31}P spectrum at 121 MHz
subsequently obtained on a supercon; the sensitivity is much
better than previously seen. The linewidths in Hertz are
somewhat larger at the higher field for two reasons: (1) the
chemical shift anisotropy contribution is larger; (2) the
"motional narrowing limit" is not so closely approached.

    From these spectra it has been possible to measure
accurately the isotropic ^{31}P chemical shift of hydroxyapatite.
The value obtained agreed with a subsequent measurement using
^{31}P MAS-NMR techniques.[6] When diphosphonate in excess of the
amount adsorbed was added to the hydroxyapatite suspension, a
high-resolution ^{31}P NMR peak from the diphosphonate dissolved
in solution was observed superimposed on the broader peak arising
from the surface diphosphonate. This result indicated that the
surface diphosphonate was <u>not</u> exchanging rapidly with the
diphosphonate in solution (on a time scale less than about
10^{-2}s), but was instead tightly bound. The surface area
occupied by the bound diphosphonate was estimated to be about

$36Å^2$ per molecule. In this example, no significant change in the ^{31}P chemical shift of the diphosphonate upon binding was observed, precluding any conclusions about the chemical nature of the interaction.

Although these experiments established a new approach for narrowing the NMR peaks of solids and surface-adsorbed species, the method suffers some limitations. First of all, it is not generally easy to form a stable colloidal suspension of small particles. Secondly, to achieve useful narrowing the NMR peak of the solid must be fairly narrow to begin with. These limitations led to the use of the previously-developed techniques of "magic-angle spinning" (MAS) and cross-polarization (CP) from protons to characterize calcium phosphates by ^{31}P CP-MAS-NMR.[6] However, while this type of ^{31}P CP-MAS-NMR is very useful for characterizing bulk calcium phosphates, it is less likely to prove as useful in characterizing their surfaces, since the ^{31}P signal from the surface will generally be overshadowed by the signal from the underlying solid. Also, the ^{31}P MAS-NMR spectra of hydroxyapatite and fluorapatite are virtually identical[6] (although differing in their cross-polarization properties). Thus, another NMR method which probes the fluoride environment more directly would be preferable for studying the surface fluoride chemistry of hydroxyapatite. ^{19}F MAS-NMR is such a method and is discussed in the remainder of this paper.

MAGIC-ANGLE SPINNING (MAS) NMR

Background

We have seen in the previous section how the anisotropic interactions of nuclear spins in a solid, which greatly broaden the NMR spectrum, can be averaged away by the random tumbling motions of a colloidal particle in suspension. A similar sort of averaging effect can be achieved by mechanically rotating a solid sample at high speed. The effect of such spinning upon the interactions contributing to an NMR spectrum is indicated in Figure 3. When the axis of rotation makes an angle of 54.7°, the "magic-angle", with respect to the main magnetic field, the anisotropic interactions are averaged to zero. The result is a high-resolution NMR spectrum characterized by isotropic chemical shifts (and any resolved spin-spin "J" couplings that are not decoupled). The ability to resolve groups with different chemical shifts has made MAS-NMR of solids a powerful technique.

One of the first nuclei to be studied using MAS-NMR was ^{19}F, which has a natural abundance of 100% and a very favorable inherent sensitivity.[7] However, in order to average to zero the generally large $^{19}F-^{19}F$ dipolar couplings, it is

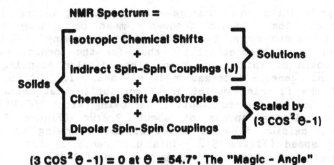

Fig. 3. NMR interactions influencing the spectrum of a spin-1/2 nucleus, and the effect of rapid spinning of a solid sample about an axis at angle θ with respect to the main magnetic field.

necessary to spin the sample at a frequency greater than the spectral width due to the dipolar interactions. Since these dipolar interactions are typically many kilohertz, spinning sufficiently rapidly is a technically difficult matter. Even at very high spinning rates (6.6 kHz), only partial narrowing was seen in the ^{19}F MAS-NMR spectrum of calcium fluoride.[8] Although the combination of MAS-NMR with multiple-pulse sequences which eliminate the ^{19}F-^{19}F dipolar couplings has been demonstrated to result in increased narrowing,[9] this method has not yet been widely applied.

In recent years much progress has been made in understanding the effects of magic-angle spinning at rotation frequencies which do <u>not</u> exceed the spectral width of the non-spinning sample.[10] In particular, it has been shown that MAS-NMR at such "moderate" spinning speeds will often result in a sharp central peak flanked by many sidebands spaced at multiples of the spinning frequency (see Figure 1). This occurs whenever the dominant interaction responsible for the static linewidth is inhomogeneous in the NMR sense. Examples of such inhomogeneous broadening interactions are the chemical shift anisotropy and heteronuclear dipolar couplings between isolated unlike spins. In these cases, the chemical shift anisotropy and/or internuclear distances can be obtained from an analysis of the sideband intensity patterns in MAS-NMR spectra.[10,11] Thus, the sidebands in MAS-NMR spectra of solids can provide valuable information that cannot be obtained from spectra of solutions. We will see examples of the utility of sideband intensities in subsequent sections.

Generally, the presence of strong <u>homonuclear</u> dipolar

couplings results in a homogeneous broadening interaction. In
such cases, the rotation frequency must be greater than the
static linewidth to achieve substantial narrowing. However, it
has been shown theoretically[10] that for the special case of a
linear chain of spins, the homonuclear dipolar couplings behave
as an inhomogeneous interaction in MAS-NMR. It is this arrangement of the fluorine nuclei in fluorapatite that helps make it
possible to achieve substantial narrowing with ^{19}F MAS-NMR at
moderate spinning speeds of about 3.8kHz (Figure 4). In
contrast, calcium fluoride shows little narrowing at the same
spinning speed (Figure 5). This difference in behavior under
magic-angle spinning has proven very useful in the study of
fluoridated hydroxyapatite surfaces.

Fluorohydroxyapatite solid solutions

We will now discuss ^{19}F MAS-NMR results from fluorohydroxyapatite solid solutions. These results provide information on

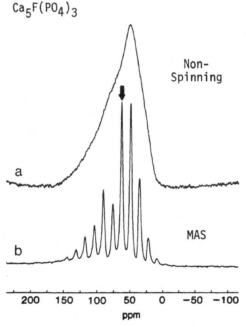

$Ca_5F(PO_4)_3$

Non-Spinning

a

MAS

b

200   150   100   50   0   -50   -100
ppm

Fig. 4.   ^{19}F NMR spectra at 282.3 MHz of synthetic fluor-
          apatite: (a) non-spinning; a large chemical shift
          anisotropy is the dominant source of broadening (see
          text); (b) with magic-angle spinning (MAS) at 3.80 kHz,
          line-broadening = 141 Hz, 30° pulses at 1s intervals,
          1000 scans, arrow indicates center peak at 64.0 ppm.

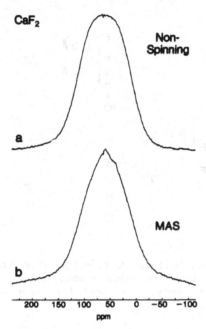

Fig. 5. ^{19}F NMR spectra at 282.3 MHz of calcium fluoride:
(a) non-spinning; homonuclear dipolar couplings are the
dominant source of broadening;(b) with magic-angle
spinning (MAS) at 3.8 kHz. A 1 μs pulse followed by a 7
μs delay before acquisition was used, and a line-
broadening of 282 Hz was applied.

how the different fluoride environments affect NMR parameters
such as isotropic chemical shifts and chemical shift anisotropies. They are also valuable in the interpretation of spectra
from fluoride-treated hydroxyapatite to be discussed in the last
section.

The general compositional formula for these solid solutions
is $Ca_5F_x(OH)_{1-x}(PO_4)_3$,   with   the   end   members   being   the
isostructural compounds hydroxyapatite and fluorapatite.[12] In
fluorapatite the fluorine atoms form an infinite linear chain
along the hexagonal axis (c-axis). Each fluorine atom has two
equally distant fluorine neighbors and is located in the center
of a triangle of calcium atoms. In hydroxyapatite, the hydroxyl
groups are aligned along the c-axis, and slightly out of the
plane of the calcium triangle.[13]

Crystallographic information on the fluorohydroxyapatite
solid solutions has been limited to the measurement of a

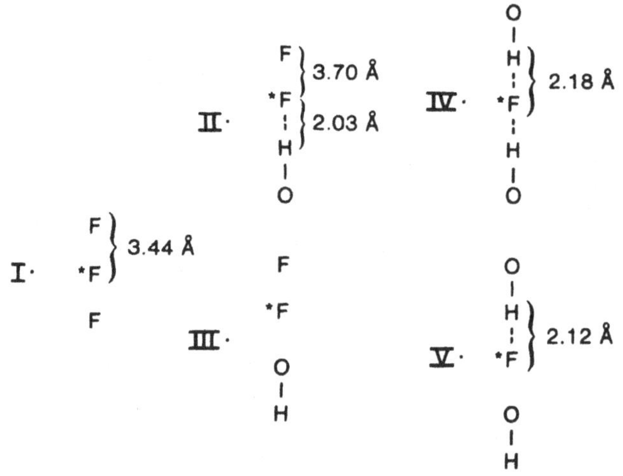

Fig. 6   Configurations about fluoride atom (with asterisk) along
         c-axis in fluorohydroxyapatites.   Dashed lines are
         hydrogen bonds.   See text for references for interatomic
         distances.

contraction in the a-axis dimension with increasing fluoride
level using X-ray powder diffraction.[12]   More-detailed
structural information about the configuration of atoms along the
c-axis chain has come from ^{1}H and ^{19}F NMR studies of single
crystals of both mineral hydroxyapatite containing fluoride
impurities[14,15] and mineral fluorapatite containing hydroxyl
impurities.[16,17]   Figure 6 shows the configurations observed in
these single crystals along with some interatomic distances
derived from the observed [19]F-^{1}H dipolar couplings.   A
fluoride ion can form a hydrogen bond with a hydroxyl hydrogen,
resulting in a bond distance shorter than would be observed if
both groups occupied their normal unperturbed positions.   This
hydrogen bond results in a lowering of the OH stretching and
librational frequencies, as seen in the IR spectra of powdered
fluorohydroxyapatite samples prepared hydrothermally.[18]

        The fluorohydroxyapatite samples we investigated were
prepared by aqueous precipitation.   IR spectra in the OH
stretching and librational regions agreed well with the spectra
obtained from hydrothermal samples,[18] interpolating spectra to
take into account the differences in exact degree of
substitution.   The IR results support the solid-solution
character of these samples, with no substantial degree of

immiscibility (segregated regions of fluorapatite and hydroxyapatite).

The MAS-NMR results also support this conclusion. The ^{19}F MAS-NMR spectra of the three fluorohydroxyapatite samples with 81%, 41%, and 24% of the maximal fluoride substitution possible are shown in Figure 7, with the spectrum of fluorapatite for comparison. The isotropic chemical shifts are listed in Table 1 with those of other related compounds.[19] Significant differences between the samples are readily evident by ^{19}F MAS-NMR. The isotropic chemical shift gradually moves upfield by about 3 ppm as the fluoride level decreases. However, the resolution is not sufficient ($\Delta\nu_{1/2} \sim 4$ ppm) to observe separate resonances from the individual chain configurations represented in Figure 6.

For the sample with 81% of maximal fluoride substitution

Table 1. ^{19}F NMR Isotropic Chemical Shifts of Fluorides

Compound	^{19}F Isotropic Chemical Shift[a]
$Ca_5F(PO_4)_3$	64.0 ppm[b]
$Ca_5F_{0.81}(OH)_{0.19}(PO_4)_3$	63.7 ppm[b]
$Ca_5F_{0.41}(OH)_{0.59}(PO_4)_3$	61.7 ppm[b]
$Ca_5F_{0.24}(OH)_{0.76}(PO_4)_3$	60.9 ppm[b]
CaF_2	56.6 ppm[c]
BaF_2	153.5 ppm[c]
MgF_2	-28 ppm[d]
NaF	-62.0 ppm[c]
KF	36.6 ppm[c]

a. relative to C_6F_6.
b. ^{19}F MAS-NMR (this work), estimated relative accuracy ±0.2 ppm, spherical bulb of C_6F_6 used as reference, no corrections for sample susceptibility.
c. multiple-pulse ^{19}F NMR, taken from reference (19).
d. multiple-pulse ^{19}F NMR, taken from reference (21).

$$Ca_5F_X(OH)_{1-X}(PO_4)_3$$

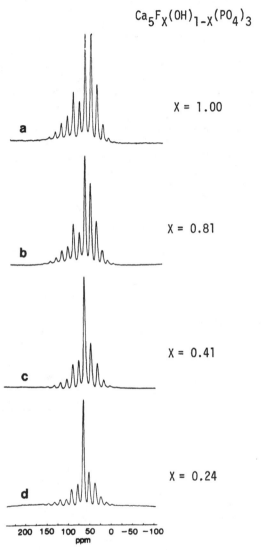

X = 1.00

a

X = 0.81

b

X = 0.41

c

X = 0.24

d

200 150 100 50 0 −50 −100
ppm

Fig. 7. ^{19}F MAS-NMR spectra at 282.3 MHz of fluorapatite and fluorohydroxyapatites of indicated composition, obtained with 30° pulses under non-saturating conditions, line-broadening = 141 Hz: (a) spinning rate = 3.80 kHz; (b) spinning rate = 3.66 kHz; (c) spinning rate = 4.00 kHz; (d) spinning rate = 3.8 kHz.

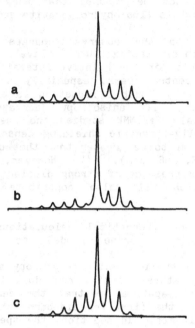

Fig. 8.   Theoretical ^{19}F MAS-NMR spectra at 282.3 MHz for an isolated ^{19}F-^{1}H group in fluorohydroxyapatite. An internuclear distance of 2.12Å and an axially-symmetric chemical-shielding tensor with $\sigma_{||}$ along the internuclear axis is assumed. The assumed spinning rate is 4.00 kHz, a line-broadening of 4.0 ppm has been applied, and the spectral width is 64 kHz. Three different values of the chemical-shift anisotropy were assumed: (a) 82 ppm; (b) 56 ppm; (c) 30 ppm. See text for acknowledgement.

(x=.81), configurations I-III are probably present on statistical grounds.   For the sample with x=.41, alternating fluoride and hydroxyl groups are most probable statistically.   In this case, configuration V is most likely to occur, since configuration IV, with two hydrogen bonds per fluorine, would deprive alternate fluorine atoms of the possibility of hydrogen bonding. For the sample with x=.24, configurations IV and V are statistically most probable. Although these configurations have been observed in the impure hydroxyapatite single-crystal NMR study,[14,15] it cannot be assumed that they are found in the same proportion in this fluorohydroxyapatite sample.

In addition to the observed changes in the isotropic chemical shift with decreasing fluoride levels, there is a marked progressive reduction in the relative intensity of the sidebands relative to the central peak, especially noticeable with the first right sideband. Such a reduction is expected in MAS-NMR when the chemical shift anisotropy decreases.[10,11] Indeed, from single-crystal ^{19}F NMR studies the chemical-shift anisotropy of the axially-symmetric shielding tensor of fluorapatite (84 ppm)[20] is known to be greater than that of configuration II in Figure 6 (35, 56 ppm).[16,17] However, the situation is complicated by the presence of strong dipolar coupling to one or two protons, which will also contribute to the sideband intensity.

The results of theoretical calculations of ^{19}F MAS-NMR line-shapes for a fluorine bonded to a single proton (configuration V in Figure 6) for three assumed values of the (axially-symmetric) chemical-shift anisotropy are shown in Figure 8. Qualitatively, these spectra reproduce the trend observed in the fluorohydroxyapatite spectra: the center peak is more intense relative to the first few sidebands than in the spectrum of fluorapatite itself. In addition, the spectrum of the x=.41 fluorohydroxyapatite sample resembles the theoretical spectrum 8c. The greater intensity of the center peak could be reproduced in the theoretical spectrum by slightly increasing the assumed chemical-shift anisotropy. As discussed earlier, this sample is the one most likely to have only one important chain configuration (V). Although we have not made detailed fits of theoretical and experimental spectra, these results suggest that the ^{19}F chemical shift anisotropies for the x=.41 and .24 samples are slightly greater than 30 ppm, but significantly less than that in pure fluorapatite (84 ppm).

Although some success has been achieved in correlating ^{19}F isotropic chemical shifts with cation electronegativity for metal fluorides with no chemical shift anisotropy,[21] interpretation of the isotropic shifts of the compounds studied here is made difficult by the different cation geometry about the fluoride. It is noteworthy that the change in the chemical shift anisotropies in the fluorohydroxyapatites appears to be about an order of magnitude larger than the change in isotropic shifts. A similar observation applies to the effects of protonation upon ^{31}P shift tensors in phosphates,[6] and demonstrates the need to consider the anisotropic nature of chemical shifts in any attempted explanation. For the present, we will use the isotropic shifts and observed sideband intensities empirically as "fingerprints" of the compounds.

^{19}F MAS-NMR Investigation of Hydroxyapatite Surfaces Exposed to Fluoride Ion

The surface of hydroxyapatite has a remarkable avidity for the fluoride ion in aqueous solution. Fluoride in solution at the part per million (50 μM) level is readily taken up.[22] The details of this process depend upon factors such as solution fluoride, calcium, and phosphate concentrations, pH, exposure time, temperature, and the nature of the hydroxyapatite surface. The last factor will be influenced by the wide variations encountered in hydroxyapatite samples in properties such as Ca/P ratio, substitutional impurities such as carbonate and sodium, hydroxyl group vacancies, etc.. Further confounding attempts to study this process are the inherent limitations in the experimental techniques that have been employed. Nevertheless, because an understanding of the process is so important in developing a chemically-sound explanation of the well-known anti-caries[22] effect of fluoride ion, many studies have been carried out.[22] The role of fluoride in bone has also been extensively studied,[23] but few surface-chemical investigations have been carried out.

We will not attempt to summarize, much less critically evaluate, past investigations here. Instead, we will list various conceivable modes of fluoridation of hydroxyapatite:

1) an ion-exchange mechanism where one F^- replaces one OH^- of hydroxyapatite at the surface;

2) an ion-exchange mechanism where one F^- replaces a phosphate group of hydroxyapatite at the surface, with compensating counter-ion movement;

3) an adsorption mechanism (chemisorption) where F^- occupies a hydroxyl site at the surface, possibly forming a hydrogen bond to the neighboring hydroxyl as is observed with fluorohydroxyapatites;

4) an adsorption mechanism with F^- at a site other than the hydroxyl site;

5) crystal growth of a bulk-phase of fluorapatite, fluorohydroxyapatite, or calcium fluoride either on the hydroxyapatite surface or (less likely) physically distinct;

6) growth of amorphous phases.

Modes 1), 3), and 5) have been implicated in various studies cited below.

How well do existing experimental methods discriminate among the above possibilities? X-ray powder diffraction is capable of identifying only bulk phases as in 5) with limited sensitivity and ability to discriminate between fluorapatite and hydroxyapatite.[24-27] Grazing-angle electron diffraction can identify thin (15-30Å) crystalline films of calcium fluoride on surfaces,[28] but cannot distinguish fluoroapatite from hydroxyapatite. Infrared spectroscopy is useful for observing the interaction between OH^- and F^-, either at the surface[26,29] or in mixed fluorohydroxyapatites,[18] but not for detecting small levels of fluorapatite or calcium fluoride in the presence of hydroxyapatite. X-ray photoelectron spectroscopy (XPS, or ESCA) is a technique useful for measuring elemental ratios of atoms at the surface (e.g. Ca/P/F), and it does establish the fact that fluoride is at the surface.[30-32] However, it cannot reliably distinguish the different forms of fluoride based on their spectral positions (chemical shifts), and the elemental ratios themselves are averages over a possibly heterogeneous sample. Electrophoretic measurements are useful for studying changes in surface charges caused by fluoride, but their interpretation is not without ambiguities, and quantitation of components is not feasible.[31,33]

In terms of chemical analytical methods, rather than physical methods, the measurement of relevant ion concentrations during and after the fluoridation process provides useful information,[34-36] but is limited by experimental inaccuracies. A widely-used analytical approach to distinguish between fluoroapatite and calcium fluoride has been developed based on the concentration of fluoride in solid samples before and after extraction with 1M KOH presaturated with fluorapatite.[37] The calcium fluoride is dissolved preferentially. However, the behavior of surface fluoride upon extraction is not known.

The principal advantages of the ^{19}F MAS-NMR method are:

1)   it selectively probes only the fluoride environment;

2)   it detects all the fluoride present, whether crystalline, amorphous, or adsorbed, in a quantitative fashion;

3)   low levels of fluoride (<0.1%) can be detected in solid samples, possibly including actual dental enamel;

4)   many NMR parameters can be used to characterize the samples.

In our study, we are investigating the effects of the solution fluoride concentration upon the form in which fluoride

occurs at the surface. Figure 9 shows the amount of fluoride
deposited after one hour of treatment at an initial pH of 7.0 vs.
the final fluoride solution concentration. The curve resembles
in shape that obtained by other workers,[31] but the fluoride
deposited per unit surface area in the plateau region is about
twice as much. The concentration range of this plateau region
approximates the fluoride concentrations to which tooth enamel is
exposed when a sodium fluoride dentifrice or mouthrinse is
used. The increase in slope beyond the plateau may be related to
the deposition of calcium fluoride at higher fluoride levels.[31]
The amount of fluoride deposited in this region increases with
longer reaction times, whereas in the plateau region the amount
deposited changes very little after the first few minutes of
reaction. We are currently obtaining uptake curves using pH-
statted samples of pre-equilibrated hydroxyapatite while
monitoring all ion concentrations; however, Figure 9 represents
well the general features of the uptake.

Since our ^{19}F MAS-NMR experiments on fluoridated
hydroxyapatite samples are ongoing, we will not try to offer a
comprehensive explanation of the fluoridation process. Rather,
we will discuss answers to several questions we have posed:

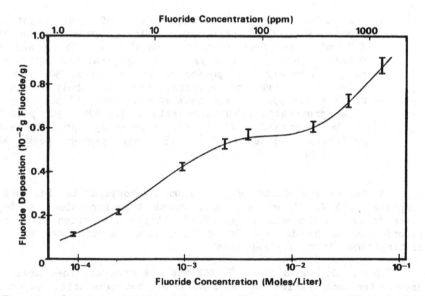

Fig. 9. Uptake of fluoride ion from aqueous solution onto
         hydroxyapatite with stirring at pH 7.0 and 37°C vs.
         final fluoride concentration. See Experimental section
         for details.

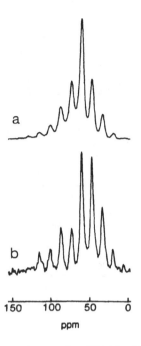

Fig. 10. ^{19}F MAS-NMR spectra at 282.3 MHz of hydroxyapatite
exposed without stirring to a (final) concentration of
9.7 mM (184 ppm) fluoride, total fluoride uptake =
0.68%: (a) several days after preparation, spinning
rate = 3.86 kHz, 30° pulses at 1s intervals, 300 scans,
80Hz line-broadening applied, real half-height linewidth = 5.9 ppm, center peak at 62.6 ppm; (b) six months
after preparation, spinning rate = 3.8 kHz, 30° pulses
at 1s intervals, 141 Hz line-broadening applied, real
half-height linewidth = 4.5 ppm, center peak at
ca. 63 ppm.

(1) Is there any evidence for fluorohydroxyapatite in these
samples? (2) Are there any solid-state transformations taking
place in isolated powders? (3) Can multiple adsorption sites of
fluoride be observed? (4) Can calcium fluoride be observed and
distinguished from fluorapatite?

Figure 10a shows the ^{19}F MAS-NMR spectrum obtained several
days after preparation of a sample of hydroxyapatite exposed
without stirring to a (final) concentration of 184 ppm F$^-$
(total fluoride uptake = 0.68%). The sideband intensities and
isotropic chemical shift (62.6 ppm) correspond to those expected
for a fluorohydroxyapatite with .41< x <.81, rather than

fluorapatite (cf . Figure 7). Using the percentage of fluoride present and the specific surface area of the sample, we calculate that there are 1.1 fluoride ions per surface hydroxyl site (assuming two hydroxyl sites per $65\AA^2$, the area of an ac or bc face;[13] the assumed value of one hydroxyl site per $76\AA^2$ used in reference (31) is inappropriate, since it refers to the ab face, which is a minor face).

The ^{19}F MAS-NMR spectrum of the same powdered sample after storage at room temperature for six months is shown in Figure 10b. This spectrum, with its increased sideband intensities, is characteristic of fluorapatite (Figure 4b). Thus, ^{19}F MAS-NMR has demonstrated the existence of a slow solid-state transformation at the surface from fluorohydroxyapatite to fluorapatite, perhaps mediated by surface-adsorbed water. We have seen similar changes with time in many samples, a sharpening of the peaks being accompanied by an increase in sideband intensities.

It is interesting to note that a single monolayer of fluoride on the ac or bc faces of hydroxyapatite would be expected to yield a ^{19}F MAS-NMR signal characteristic of fluorapatite, since each fluorine would presumably be in the center of a triangle of calcium atoms and have two fluorine nearest-neighbors. However, a single monolayer of fluoride on the ab face would result in each fluorine having one hydroxyl neighbor; the resulting NMR signal might resemble that of a fluorohydroxyapatite. Unfortunately, it would appear difficult to use ^{19}F MAS-NMR to detect differences in the reactivity of the different faces to fluoride, since less than a monolayer of fluoride on the ac or bc faces would also yield a spectrum characteristic of a fluorohydroxyapatite.

The effect of the aqueous fluoride concentration on the form in which fluoride occurs in samples aged for many months after isolation is shown in Figure 11. Figure 11b is the ^{19}F MAS-NMR spectrum of a lower concentration sample (exposed to a final fluoride concentration of 184 ppm), and shows only the presence of fluorapatite. In marked contrast, the sample exposed to a much higher fluoride concentration (2950 ppm) exhibits a sharp peak with sidebands characteristic of fluorapatite as well as an underlying broad peak (Figure 11a). We assign the broad peak to calcium fluoride on the basis of its chemical shift position and large linewidth, as well as its chemical plausibility. The ^{19}F MAS-NMR data thus provide spectroscopic confirmation of the onset of calcium fluoride formation at higher concentrations.[31,34] Investigation of samples exposed to intermediate concentrations suggests that the increase in slope beyond the plateau region of the fluoride uptake curve (Figure 9) coincides with the appearance of a broad component in the ^{19}F MAS-NMR spectra. Although

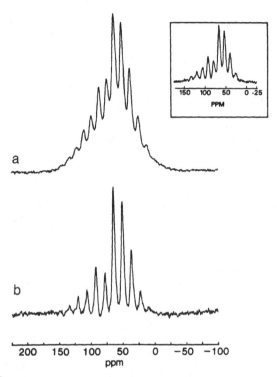

Fig. 11. ^{19}F MAS-NMR spectra at 282.3 MHz of hydroxyapatite
exposed without stirring to fluoride, aged 14 (a) and 10
(b) months after preparation: (a) 155.3 mM (2950 ppm)
final fluoride concentration, total fluoride uptake =
2.2%, spinning rate = 3.80 kHz, 45° pulses at 1s
intervals; Insert shows spectrum of same sample obtained
using Hahn spin echo (90_x - 263µs - 180_y - 263µs -
acquire), 90° pulse = 3.2 µs; (b) 9.7 mM (184 ppm) final
fluoride concentration, total fluoride uptake = 0.68%,
spinning rate = 3.84 kHz, 45° pulses at 1s intervals.
Center peak of fluorapatite component in both samples
is at 63.3 ppm.

this broad peak is presumed to be calcium fluoride, its low
relative intensity makes an unequivocal assignment difficult.

Identification and quantitation of . the fluorapatite
component in a spectrum such as that in Figure 11a is hindered by
the substantial broad peak arising from calcium fluoride. It is
possible to eliminate the calcium fluoride signal from the

spectrum by taking advantage of the fact that the spin-spin relaxation time T_2 of calcium fluoride is two orders of magnitude smaller than that of fluorapatite.[3] The insert in Figure 11a shows a ^{19}F MAS–NMR spectrum of the same sample obtained by using a Hahn spin-echo pulse sequence[38] (the delay time must be set to an integral multiple of one sample rotation period to obtain a good spectrum[39]). All of the signal from calcium fluoride has decayed, leaving only the signal from the fluorapatite component, slightly reduced in intensity. This spectrum closely resembles that of pure fluorapatite rather than those of fluorohydroxyapatites (Figure 7). Integration of such a spectrum can be used to quantitate the apatitic component in treated samples.

We note that it should be possible to obtain the spectrum of the non-apatitic component by subtracting the Hahn spin-echo spectrum from the single-pulse spectrum (Figure 11a) using the appropriate scaling. Difficulties in correctly phasing the two spectra make it difficult to obtain a reliable difference spectrum. We are working to overcome these problems so that the Hahn spin-echo sequence can be used to quantitatively measure the relative amounts of calcium fluoride and fluorapatite in surface samples, using synthetic mixtures as standards.

We believe it is highly significant that all of the (aged) fluoride-treated hydroxyapatite samples we have investigated have exhibited sharp ^{19}F MAS–NMR signals arising from fluorapatite or a fluorohydroxyapatite-like component, but no other sharp peaks. This observation demonstrates that fluoride ion occupies the hydroxyl sites in the apatite lattice. If fluoride were present at some other side, relatively distant from other ions, then one would expect to see a sharp peak at a different chemical shift and with a different sideband intensity pattern. It can also be concluded that fluoride ions occupying the hydroxyl sites are surrounded by all three calcium atoms, since if one or two calcium atoms were missing from the triangle, one would expect significant changes in the chemical shift and relative sideband intensities.

CONCLUSIONS

By preparing a colloidal suspension of very small particles, it is possible to obtain relatively high-resolution NMR spectra of solids and surface-adsorbed species for selected systems using only a conventional high-resolution NMR spectrometer. The ^{31}P NMR studies of the diphosphonate $Na_2(CH_3)C(OH)(PO_3H)_2$ adsorbed to the surface of hydroxyapatite reveal that it is tightly bound to the surface and does not exchange rapidly with the free diphosphonate in solution. The amount of hydroxyapatite surface area per tightly bound diphosphonate can be measured.

Another useful NMR technique for studying surfaces is magic-angle spinning (MAS) NMR. ^{19}F MAS-NMR has great utility in exploring the surface chemistry of the reaction of fluoride ion with hydroxyapatite. In particular, we have shown that:

1) Calcium fluoride, fluorapatite, and fluorohydroxy-apatite solid solutions exhibit widely different and characteristic spectra because of the different NMR interactions present in each case.

2) In the hydroxyapatite samples exposed to fluoride at pH 7, the first form of fluoride resembles a fluorohydroxyapatite with 40 to 80% substitution of fluoride for hydroxide ions.

3) These fluoride-treated samples undergo a solid-state transformation over several months which results in the appearance of a fluorapatite spectrum with sharper peaks.

4) No evidence for multiple fluoride adsorption sites has been seen so far. The fluoride ion appears to occupy its normal apatitic position, surrounded by a triangle of calcium atoms.

5) Calcium fluoride, as well as fluoroapatite, is seen in aged hydroxyapatite samples exposed to high fluoride concentrations at neutral pH. In aged samples exposed to lower concentrations of fluoride, only fluorapatite is observed.

6) Quantitative analysis for the amounts of calcium fluoride and fluorapatite in a sample should be possible by making use of the large difference in their T_2 relaxation times.

Further ^{19}F NMR studies should be capable of providing information concerning the motions of the fluoride ion at the surface, and about possible contact at the atomic level between fluorapatite and calcium fluoride phases. Another approach to the characterization of the surface of fluoride-treated hydroxy-apatite which we have recently developed is ^{19}F-^{31}P cross-polarization (CP) MAS-NMR. The ^{31}P NMR signal in this experiment arises solely from those phosphorus atoms near fluorine atoms, and thus is a complementary probe of the hydroxyapatite surface. In summary, high-resolution NMR of solids should continue to provide further detailed information about the surface of this important and complex biological mineral.

ACKNOWLEDGEMENTS

We would like to thank Mr. Robert Faller of the Sharon Woods Technical Center for his technical assistance with the preparation of the surface fluoride samples. We also wish to thank Dr. Donald White of the Sharon Woods Technical Center for his contributions to the further characterization of the reaction of fluoride with hydroxyapatite.

REFERENCES

1.   T. M. Duncan and C. Dybowski, Chemisorption and surfaces studied by nuclear magnetic resonance spectroscopy, Surface Science Reports, 1:157 (1981).

2.   J. P. Yesinowski, High-resolution NMR spectroscopy of solids and surface-adsorbed species in colloidal suspension: ^{31}P NMR spectra of hydroxyapatite and diphosphonates, J. Am. Chem. Soc., 103:6266 (1981).

3.   J. P. Yesinowski, R. A. Wolfgang, and M. J. Mobley, ^{19}F MAS-NMR of fluorapatite, fluoro-hydroxyapatite solid solutions and related compounds, abstract and poster presented at 23rd Experimental NMR Conference, Madison, April, 1982; J. P. Yesinowski, manuscript in preparation.

4.   J. P. Yesinowski and M. J. Mobley, ^{19}F MAS-NMR of fluoridated hydroxyapatite surfaces, manuscript in preparation.

5.   H. G. McCann, The solubility of fluorapatite and its relationship to that of calcium fluoride, Archs. oral Biol., 13:987 (1968).

6.   W. P. Rothwell, J. S. Waugh, and J. P. Yesinowski, High-resolution variable-temperature ^{31}P NMR of solid calcium phosphates, J. Am. Chem. Soc., 102:2637 (1980).

7.   E. R. Andrew, in "Prog. in NMR Spectroscopy", eds. J. W. Emsley, J. Feeney, and L. H. Sutcliffe, 8 Part 1, 1-39 (1971), and references therein.

8.   I. J. Lowe, Free induction decays of rotating solids, Phys. Rev. Letters, 2:285 (1959).

9.   B. C. Gerstein, R. G. Pembleton, R. C. Wilson, and L. M. Ryan, High resolution NMR in randomly oriented solids with homonuclear dipolar broadening: combined multiple pulse NMR and magic angle spinning, J. Chem. Phys., 66:361 (1977).

10. M. M. Maricq and J. S. Waugh, NMR in rotating solids, J. Chem. Phys., 70:3300 (1979).

11. J. Herzfeld and A. E. Berger, Sideband intensities in NMR spectra of samples spinning at the magic angle, J. Chem. Phys., 73:6021 (1980).

12. E. C. Moreno, M. Kresak, and R. T. Zahradnik, Physico-chemical aspects of fluoride-apatite systems relevant to the study of dental caries, Caries Res., Suppl. 1, 11:142 (1977).

13. M. I. Kay, R. A. Young, and A. S. Posner, Crystal structure of hydroxyapatite, Nature, 204:1050 (1964).

14. R. A. Young, W. van der Lugt, and J. C. Elliott, Mechanism for fluorine inhibition of diffusion in hydroxyapatite, Nature, 223:729 (1969).

15. W. van der Lugt, D. I. M. Knotterus, and W. G. Perdok, Nuclear magnetic resonance investigation of fluoride ions in hydroxyapatite, Acta. Cryst., B27:1509 (1971).

16. R. G. Knoubovets, M. L. Afanasjev, and S. P. Habuda, Hydrogen bond and ^{19}F NMR chemical shift anisotropy in apatite, Spectroscopy Letters, 2:121 (1969).

17. A. M. Vakhrameev, S. P. Gabuda, and R. G. Knubovets, ^1H and ^{19}F NMR in apatites of the type $Ca_5(PO_4)_3[F_{1-x}(OH_x)]$, J. Struct. Chem. (USSR), 19:256 (1978).

18. F. Freund and R. M. Knobel, Distribution of fluorine in hydroxyapatite studied by infrared spectroscopy, J. Chem. Soc. Dalton, 1136 (1977).

19. D. P. Burum, D. D. Elleman, and W.-K. Rhim, A multiple pulse zero crossing NMR technique, and its application to ^{19}F chemical shift measurements in solids, J. Chem. Phys., 68:1164 (1978).

20. J. L. Carolan, A pulsed NMR investigation of ^{19}F chemical shift anisotropy in single crystals of fluoro-apatite, Chem. Phys. Letters, 12:389 (1971).

21. R. W. Vaughan, D. D. Elleman, W.-K. Rhim, and L. M. Stacey, ^{19}F chemical shift tensor in group II difluorides, J. Chem. Phys., 57:5383 (1972).

22. W. E. Brown and K. G. Konig, eds., Caries Res., 11, Suppl. 1, 1-327 (1977), and references therein.

23. E. D. Eanes and A. H. Reddi, The effect of fluoride on bone mineral apatite, Metab. Bone Dis. & Rel. Res., 2:3 (1979).

24. S. H. Y. Wei and W. C. Forbes, X-ray diffraction analyses of the reactions between intact and powdered enamel and several fluoride solutions, J. Dent. Res., 47:471 (1968).

25. C. A. Baud and S. Bang, Electron probe and X-ray diffraction microanalyses of human enamel treated in vitro by fluoride solution, Caries Res., 4:1 (1970).

26. E. J. Duff, An infrared and X-ray diffractometric study of the incorporation of fluoride into hydroxyapatite under conditions of the cyclic variation of pH, Archs. oral Biol., 11:763 (1975).

27. B. Laufer, I. Mayer, I. Gedalia, D. Deutsch, H. W. Kaufman, and M. Tal, Fluoride-uptake and fluoride-residual of fluoride-treated human root dentine in vitro determined

by chemical, scanning electron microscopy and X-ray diffraction analyses, Archs. oral Biol. , 26:159 (1981).

28. M. D. Francis, J. A. Gray, and W. J. Griebstein, The formation and influence of surface phases on calcium phosphate solids, Adv. in Oral Biology, 3:83 (1968).

29. B. Menzel and C. H. Amberg, An infrared study of the hydroxyl groups in a nonstoichiometric calcium hydroxyapatite with and without fluoridation, J. Colloid Interf. Sci., 38:256 (1972).

30. D. M. Hercules and N. L. Craig, Composition of fluoridated dental enamel studied by X-ray photoelectron spectroscopy (ESCA), J. Dent. Res., 55:829 (1976).

31. J. Lin, S. Raghavan, and D. W. Fuerstenau, The adsorption of fluoride ions by hydroxyapatite from aqueous solution, Colloids and Surfaces, 3:357 (1981).

32. H. Uchtmann and H. Duschner, Electron spectroscopic studies of interactions between superficially-applied fluorides and surface enamel, J. Dent. Res., 61: 423 (1982).

33. S. Chander, C. C. Chiao, and D. W. Fuerstenau, Transformation of calcium fluoride for caries prevention, J. Dent. Res., 61:403 (1982).

34. H. G. McCann, Reactions of fluoride ion with hydroxyapatite, J. Biol. Chem., 201:247 (1953).

35. M. A. Spinelli, F. Brudevold, and E. Moreno, Mechanism of fluoride uptake by hydroxyapatite, Archs. oral Biol., 16:187 (1971).

36. F. F. Feagin, Calcium, phosphate, and fluoride deposition on enamel surfaces, Calc. Tiss. Res., 8:154 (1971).

37. V. Caslavska, E. C. Moreno, and F. Brudevold, Determination of the calcium fluoride formed from in vitro exposure of human enamel to fluoride solutions, Archs. oral Biol., 20:333 (1975).

38. A. Abragam, "The Principles of Nuclear Magnetism", Oxford University Press, London, pp. 33-34, 58-63, (1971).

39. W. T. Dixon, Spinning-sideband-free and spinning-sideband-only NMR spectra in spinning samples, J. Chem. Phys., 77:1800 (1982).

SUBJECT INDEX

177